Arthur Sherburne Hardy

Elements of analytic geometry

Arthur Sherburne Hardy

Elements of analytic geometry

ISBN/EAN: 9783337276379

Printed in Europe, USA, Canada, Australia, Japan

Cover: Foto ©berggeist007 / pixelio.de

More available books at **www.hansebooks.com**

ELEMENTS

OF

ANALYTIC GEOMETRY.

BY

ARTHUR SHERBURNE HARDY, Ph.D.,

Professor of Mathematics in Dartmouth College.

———•○❊○•———

BOSTON, U.S.A.:
PUBLISHED BY GINN & COMPANY.
1889.

Typography by J. S. Cushing & Co., Boston, U.S.A.

Presswork by Ginn & Co., Boston, U.S.A.

PREFACE.

ALTHOUGH writing a text-book for the use of beginners following a short course, the tendency of an author is to sacrifice the practical value of the treatise to completeness, generalization, and scientific presentation. I have endeavored to avoid this error, which renders many works unsuitable for the class-room, however valuable they may be for reference, and yet to encourage the habit of generalization. To this end I have attempted to shun the difficulties involved in introducing the beginner to the conics, before he is familiar with their forms, through the discussion of the general equation; and at the same time to secure to him the advantages of a general analysis of the equation of the second degree. The teacher will observe the same effort to cultivate the power of general reasoning, which it is one of the objects of Analytic Geometry to promote, in the preliminary construction of loci, a process too often left in the form of a merely mechanical construction of points by substitution in the equation. In passing from Geometry to Analytic Geometry, the student should see that, while the field of operations is extended, the subject matter is essentially the same; and that what is fundamentally new is the *method*, the lines and surfaces of Geometry being replaced by their equations. His chief difficulties are:

First. A thorough understanding of the device by which this substitution is effected; hence considerable attention has been paid to this simple matter.

Second. The acquisition of an independent use of the new method as an instrument of research; hence the insertion of problems illustrative of the analytic, as distinguished from the geometric, method of proof. The function of numerical examples — that is, of examples consisting of a mere substitution of numerical values for the general constants — is simply that of testing the student's knowledge of the nomenclature. The real example in Analytic Geometry is the application of the method to the discovery of geometrical properties and forms.

The polar system has been freely used. It is not briefly explained and subsequently abandoned without application; nor is it applied redundantly to what has been already treated by the rectilinear system. It is used as one of two methods, each of which has its advantages, the selection of one or the other in any given case being governed by its adaptability to the demonstration or problem in hand.

The time allotted to the courses in Analytic Geometry for which it is hoped this treatise will be found adapted has determined the exclusion of certain topics, and has limited the chapters on Solid Geometry to the elements necessary to the student in the subsequent study of Analytic Mechanics.

ARTHUR SHERBURNE HARDY.

HANOVER, N.H., Oct. 6, 1888.

CONTENTS.

PART I.—PLANE ANALYTIC GEOMETRY.

CHAPTER I.—COORDINATE SYSTEMS.

SECTION I.—THE POINT.

The Rectilinear System.

ART.		PAGE.
1.	Position of a point in a plane	1
2.	Definitions	2
3.	Construction of a point	2
4.	Definitions	2
5.	Equations of a point. Examples	3
6.	Division of a line. Examples	4
7.	Distance between two points. Examples	5

The Polar System.

8.	Position of a point in a plane	6
9.	Signs of the polar coordinates	7
10.	Construction of a point	7
11.	Equations of a point	7
12.	Definitions. Examples	8
13.	Distance between two points. Examples	8

SECTION II.—THE LINE.

The Rectilinear System.

14.	Loci, and their equations	10
15.	Distinctions between Analytic Geometry, Geometry, and Algebra	11
16.	Quantities of Analytic Geometry	12
17.	Construction of loci. Examples	13

CONTENTS.

The Polar System.

ART.		PAGE.
18.	Polar equations of loci	21
19.	Construction of polar equations	22
20.	General notation	24

SECTION III.—RELATION BETWEEN THE RECTILINEAR AND POLAR SYSTEMS.

21.	Transformation of coordinates	25

Rectilinear Transformations.

22.	Formulæ for passing from any rectilinear system to any other	27

Polar Transformations.

23.	Formulæ for passing from any rectilinear to any polar system	30
24.	Formulæ for passing from any polar to any rectilinear system	31

CHAPTER II. — EQUATION OF THE FIRST DEGREE. THE STRAIGHT LINE.

SECTION IV.—THE RECTILINEAR SYSTEM.

Equations of the Straight Line.

25.	General equation of the first degree	35
26.	Common forms	36
27.	Derivation of the common forms from the general form	38
28.	Illustrations	39
29.	Discussion of the common forms	40
30.	Construction of a straight line from its equation. Examples	43
31.	Equation of a straight line through a given point	45
32.	Equation of a straight line through two given points. Examples	45

Plane Angles.

33.	Angle between two straight lines	47
34.	Equation of a line making a given angle with a given line	48
35.	Conditions of parallelism and perpendicularity. Examples	49

Intersections.

36.	Intersection of loci. Examples	51
37.	Lines through the intersections of loci	53

CONTENTS.

Distances between Points and Lines, and Angle-Bisectors.

ART.		PAGE.
38.	Distance of a point from a line	55
39.	Another method. Examples	56
40.	Equation of the angle-bisector. Examples	58

SECTION V. — THE POLAR SYSTEM.

41.	Derivation of polar from rectangular equations	60
42.	Polar equation of a straight line. Normal form. Examples	60

SECTION VI. — APPLICATIONS.

43.	Recapitulation	63
44.	Properties of rectilinear figures	64

CHAPTER III. — EQUATION OF THE SECOND DEGREE. THE CONIC SECTIONS.

SECTION VII. — COMMON EQUATIONS OF THE CONIC SECTIONS.

45.	The Conic Sections	71

The Circle.

46.	Definitions	72
47.	General equation of the circle	72
48.	The equation of every circle some form of $y^2 + x^2 + Dy + Ex + F = 0$	73
49.	Every equation of the form $y^2 + x^2 + Dy + Ex + F = 0$ the equation of a circle	74
50.	To determine the radius and centre	75
51.	Concentric circles. Examples	75
52.	Polar equation of the circle	77

The Ellipse.

53.	Definitions	78
54.	Central equation	78
55.	Definitions	80
56.	Common form of central equation	80
57.	Length of focal radii	81
58.	Polar equation	81
59.	The ratio	82

ART.		PAGE.
60.	Geometrical construction of the ellipse	83
61.	The circle a particular case	85
62.	Varieties of the ellipse. Examples	85

The Hyperbola.

63.	Definitions	87
64.	Central equation	87
65.	Definitions	89
66.	Common form of central equation	89
67.	Length of focal radii	90
68.	Polar equation	91
69.	The ratio	92
70.	Geometrical construction of the hyperbola	93
71.	The equilateral and conjugate hyperbolas	95
72.	Varieties of the hyperbola. Examples	95

The Parabola.

73.	Definitions	98
74.	Common equation	98
75.	Polar equation	99
76.	Geometrical construction of the parabola. Examples	100

SECTION VIII.—GENERAL EQUATION OF THE CONIC SECTIONS.

77.	Definitions	102
78.	General equation of the conics. Examples	102
79.	Every equation of the second degree the equation of a conic	103
80.	Determination of species. Examples	105
81.	The equation $Ay^2 + Cx^2 + Dy + Ex + F = 0$ represents all species.	106
82.	Definitions	108
83.	Centres	108
84.	The equation $Ay^2 + Cx^2 + F = 0$ represents all ellipses and hyperbolas	108
85.	Varieties of the parabola	109
86.	Definitions	111
87.	Locus of middle points of parallel chords	111
88.	Tangents at vertices of a diameter parallel to the chords bisected by that diameter	114
89.	Definitions	114
90.	Conjugate diameters of ellipse	114
91.	Every straight line through the centre of an hyperbola meets the hyperbola or its conjugate	115

ART.		PAGE.
92.	Definitions	116
93.	Conjugate diameters of the hyperbola	116

Construction of Conics from their Equations.

94.	By comparison with the general equation. Examples	117
95.	By transformation of axes. Examples	119
96.	By conjugate diameters. Examples	122
97.	Construction where $B=0$. Examples	125

General Theorems.

98.	Conic section through five points. Examples	126
99.	Intersection of conics. Examples	128
100.	Definitions	131
101.	Conics, similar and similarly placed	131
102.	Condition for two straight lines	132

SECTION IX.—TANGENTS AND NORMALS.

103.	Definitions	134
104.	Equations of tangent and secant. Examples	134
105.	Problems and Examples	137
106.	Chord of contact	141
107.	Equation of the normal. Examples	142
108.	Definitions	143
109.	Subtangent and subnormal. Geometrical constructions	143

SECTION X.—OBLIQUE AXES.

Conjugate Diameters.

110.	Equations of ellipse and hyperbola	150
111.	Ordinates to conjugate diameters of ellipse	151
112.	Same for hyperbola	152
113.	Parameter a third proportional to the axes	152
114.	Circumscribed circle	152
115.	Sum of the squares of conjugate diameters constant	152
116.	Difference of the squares of conjugate diameters constant	153
117.	Rectangle of the focal radii	154
118.	Circumscribed parallelogram	154
119.	Equal conjugate diameters	155

Supplemental Chords.

120.	Definitions	156
121.	Property of supplemental chords. Geometrical constructions	156

Parabola referred to Oblique Axes.

ART.		PAGE.
122.	Equation of the parabola...	157
123.	Property of ordinates...	159

Asymptotes.

124.	Equation of the hyperbola...	159
125.	Property of the secant. Geometrical construction....................	161
126.	Property of the tangent. Geometrical construction...................	161
127.	Tangents meet on the asymptotes......................................	162

CHAPTER IV.—LOCI.

128.	Classification of loci..	164

SECTION XI.—LOCI OF THE FIRST AND SECOND ORDER.

129.	Examples..	166

SECTION XII.—HIGHER PLANE LOCI.

130.	1. The cardioid. Trisection of the angle............................	176
	2. The conchoid. Trisection of the angle............................	177
	3. The cissoid. Duplication of the cube.............................	179
	4. The lemniscate..	181
	5. The witch..	182
	6–8. Examples..	183

SECTION XIII.—TRANSCENDENTAL CURVES.

131.	1. The logarithmic curve..	185
	2. The cycloid..	185
	3–10. The circular functions..	187
	11. Spiral of Archimedes..	189
	12. Reciprocal spiral...	190
	13. The lituus..	191
	14. Logarithmic spiral..	192

Part II. — Solid Analytic Geometry.

Chapter V. — The Point, Straight Line, and Plane.

Section XIV. — Introductory Theorems.

ART.		PAGE.
132.	Definitions. Projections of lines	195
133.	Length of the projection of a line on a line and plane	196
134.	Projection of a broken line	197

Section XV. — The Point.

135.	Position of a point	198
136.	Equations of a point. Examples	199
137.	Distance between two points	200
138.	Polar coordinates of a point	201
139.	Relations between polar and rectangular coordinates	201
140.	Direction-angles and -cosines	202
141.	Relation between direction-cosines. Examples	202
142.	Angle between straight lines	203

Section XVI. — The Plane.

143.	General equation of a surface	205
144.	Equation of a plane	206
145.	Intercept form. Examples	207
146.	Normal form. Examples	208
147.	Equation of a plane through three points. Examples	209
148.	Angle between planes. Examples	210
149.	Traces of a plane	213

Section XVII. — The Straight Line.

150.	Equations of a straight line	214
151.	Symmetrical forms	216
152.	Reduction to symmetrical form. Examples	216
153.	Equations of a line through two points	218
154.	Angle between straight lines. Examples and problems	219

CHAPTER VI. — SURFACES OF REVOLUTION, CONIC SECTIONS, AND THE HELIX.

Section XVIII. — Surfaces of Revolution.

ART.		PAGE.
155.	Definitions	222
156.	General equation of a surface of revolution	222
157.	The sphere	223
158.	The prolate spheroid	223
159.	The oblate spheroid	224
160.	The paraboloid	224
161.	The hyperboloid of two nappes	224
162.	The hyperboloid of one nappe	225
163.	The cylinder	225
164.	The cone	226

Section XIX. — The Conic Sections.

165.	General equation of the plane section of a cone, and its discussion	227

Section XX. — The Helix.

166.	Definitions	229
167.	Equations of the helix	229

PART I.

PLANE ANALYTIC GEOMETRY.

ANALYTIC GEOMETRY.

2. Defs. The fixed lines $X'X$, $Y'Y$, are called the **axes of reference**; their intersection, O, the **origin**; the distances mP^1, nP^1, the **coordinates of the point** P^1; and to distinguish these coordinates, nP^1 is called the **abscissa**, and mP^1 the **ordinate** of P^1.

3. Construction of a point. Since the coordinates of a point, when given, fix its position with reference to the axes, and since (Fig. 1) $Om = nP^1$, $On = mP^1$, *to determine the position of a point whose coordinates are given* we have simply to lay off the given abscissa from O along $X'X$ in the direction indicated by its sign, and at its extremity on a parallel to $Y'Y$, the given ordinate, above or below $X'X$ according as it is positive or negative. This determination of the position of a point is called the *construction of the point*.

4. The axes of reference are always lettered as in Fig. 1, and hence are often designated as the axes of X and Y, the former being usually taken horizontal. For brevity they will frequently be spoken of as X and Y simply. The abscissa of a point, being always a distance parallel to the axis of X, is always represented by the letter x; and for a like reason the ordinate is always represented by the letter y; hereafter, therefore, these letters will always represent distances parallel to the axes of reference. As indicating the directions in which abscissas and ordinates are laid off, the axes are also distinguished as the axis of abscissas ($X'X$), and the axis of ordinates ($Y'Y$). The angles XOY, YOX', $X'OY'$, $Y'OX$, are known as the first, second, third, and fourth, angles, respectively.

It is evident that so long as the angle XOY is not zero, it may have any value whatever. When a right angle, the system of reference is called a **rectangular system**; otherwise, an **oblique system**. As nothing in general, is gained by assuming oblique axes, the axes will hereafter be supposed rectangular, unless mention is made to the contrary; the abscissa and ordinate of a

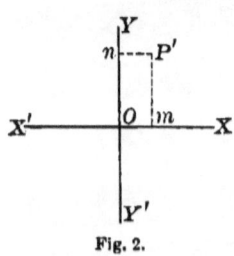
Fig. 2.

point will thus be (Fig. 2) the *perpendicular* distances of the point from the axes.

5. Equations of a point. It is now evident that we may designate the position of a point by giving its coordinates in the form of equations. Thus, $x = 2$, $y = 3$, designate a point in the first angle, distant 3 units from the axis of X, and 2 units from the axis of Y. These equations are called the **equations of the point**. But it is more usual to adopt the notation (2, 3) to designate the point, *the abscissa being always written first.*

EXAMPLES. 1. What are the signs of the coordinates of all points in the first angle? In the second? In the third? In the fourth? Where are all points situated whose ordinates are zero? Whose ordinates are equal and have the same sign? Whose abscissas are zero? What are the coordinates of the origin?

2. Construct the following points: $(2, 4)$; $(3, -2)$; $(-6, -2)$; $(-4, 3)$; $(2, 0)$; $(0, -2)$; $(-2, 0)$; $(0, 2)$; $(0, 0)$.

3. Construct the triangle whose vertices are $(4, 5)$, $(4, -5)$, $(-4, 5)$. What kind of a triangle is it, and what are the directions of its sides?

4. The side of a square is a, and its centre is taken as the origin, the axes being parallel to its sides. What are the coordinates of the vertices? What, when the axes coincide with the diagonals, the origin being still at the centre?

5. An isosceles triangle, whose base is b and altitude a, has its base coincident with X. What are the coordinates of its vertices when the origin is (1) at the centre of the base? (2) at the left hand extremity of the base?

6. Construct and name the figures whose vertices are

(1). (a, a), $(a, -a)$, $(-a, -a)$, $(-a, a)$.
(2). (a, b), $(a, -b)$, $(-a, -b)$, $(-a, b)$.
(3). (a, b), $(a, -b)$, $(-a, -b)$, $(-a, c)$.
(4). (a, b), (c, d), $(-e, d)$, $(-f, b)$.

6. *To find the coordinates of a point which divides the straight line joining two given points in a given ratio.*

Let P', P'', be the given points, x', y', and x'', y'', their coordinates, and P a third point (x, y), dividing $P'P''$ so that $P'P : PP'' :: m : n$. Then, from similar triangles,

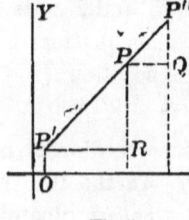

Fig. 3.

$$\frac{P'R}{PQ} = \frac{RP}{QP''} = \frac{m}{n}. \quad (1)$$

But $P'R = x - x'$, $PQ = x'' - x$, $RP = y - y'$, $QP'' = y'' - y$. Substituting these values,

$$\frac{x-x'}{x''-x} = \frac{y-y'}{y''-y} = \frac{m}{n}.$$

Solving these equations for x and y, we obtain

$$x = \frac{mx'' + nx'}{m+n}, \qquad y = \frac{my'' + ny'}{m+n}. \quad (2)$$

For the middle point of $P'P''$, $m = n$. Hence *the coordinates of the middle point of a line joining two given points are*

$$x = \frac{x'' + x'}{2}, \qquad y = \frac{y'' + y'}{2}. \quad (3)$$

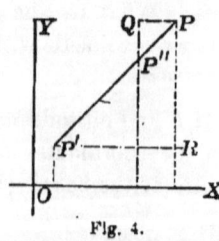

Fig. 4.

If the line is cut *externally*, we have from (1) and Fig. 4,

$$\frac{x-x'}{x-x''} = \frac{y-y'}{y-y''} = \frac{m}{n},$$

whence

$$x = \frac{mx'' - nx'}{m-n}, \qquad y = \frac{my'' - ny'}{m-n}. \quad (4)$$

OBLIQUE AXES. The above formulæ hold good for oblique axes, since the triangles remain similar whatever the angle XOY.

EXAMPLES. 1 Find the coordinates of the middle point of the line joining $(5, 3)$ and $(3, 9)$; also $(5, 8)$ and $(-5, -8)$.

Ans. $(4, 6)$; $(0, 0)$.

THE POINT.

2. Find the coordinates of the middle points of the sides of the triangle whose vertices are (6, 2), (8, 4), (10, 12).

Ans. (7, 3); (9, 8); (8, 7).

3. The line joining $(-2, -3)$ and $(-4, 5)$ is trisected. Find the coordinates of the point of trisection nearest $(-2, -3)$.

Ans. $\left(-\dfrac{8}{3}, -\dfrac{1}{3}\right)$.

4. The line whose extremities are (2, 4) and (6, −8) is divided in the ratio 3 : 2. Find the two points of division fulfilling the condition. *Ans.* $\left(\dfrac{22}{5}, -\dfrac{16}{5}\right)$; $\left(\dfrac{18}{5}, -\dfrac{4}{5}\right)$.

5. Find the two points on the line joining (2, 4) and (6, 3) twice as far from (2, 4) as from (6, 3).

Ans. (10, 2); $\left(\dfrac{14}{3}, \dfrac{10}{3}\right)$.

6. The vertices of a triangle are $(-4, -3)$, (6, 1), (4, 11). Find the coordinates of the points of trisection, farthest from the vertices, of the lines joining the vertices and the middle points of the opposite sides. *Ans.* (2, 3).

7. *To find the distance between two given points.*

Let P', P'', be the given points, x', y', and x'', y'', being their coordinates. Then $P'P'' = \sqrt{P'R^2 + RP''^2}$; or, representing $P'P''$ by d,

$$d = \sqrt{(x''-x')^2 + (y''-y')^2}. \quad (1)$$

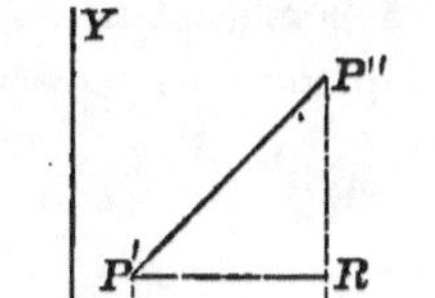

If one of the points, as P'', is at the origin, its coordinates will be zero. Hence *the distance of any point P' from the origin* is

$$d = \sqrt{x'^2 + y'^2}. \quad (2)$$

Fig. 5.

OBLIQUE AXES. In this case the triangle $P'RP''$ will not be a right triangle.
Let $\beta =$ inclination of the axes. Then

$$P'P'' = \sqrt{P'R^2 + RP''^2 - 2P'R \cdot RP'' \cos P'RP''},$$

or, since $P'RP'' = 180° - \beta$,

$$d = \sqrt{(x''-x')^2 + (y''-y')^2 + 2(x''-x')(y''-y')\cos\beta}, \quad (3)$$

which, when $\beta = 90°$, reduces to (1), since $\cos 90° = 0$.

EXAMPLES. 1. Find the distance between (2, 4) and (5, 8).

As the quantities under the radical sign are squares, it is immaterial whether we substitute (2, 4) for (x', y') and (5, 8) for (x'', y''), or vice versa. Thus
$$d = \sqrt{(2-5)^2 + (4-8)^2} = \sqrt{(5-2)^2 + (8-4)^2} = 5.$$

2. Find the distances between the following points: $(-2, -4)$ and $(-5, -8)$; $(7, -1)$ and $(-6, 1)$; $(7, 2)$ and $(-7, -2)$.

Ans. 5; $\sqrt{173}$; $2\sqrt{53}$.

3. Find the distance of $(6, -8)$ from the origin. *Ans.* 10.

4. Find the lengths of the sides of the triangle whose vertices are $(4, 8)$, $(1, 4)$, $(-4, -8)$. *Ans.* 5; 13; $8\sqrt{5}$.

5. Find the lengths of the sides of the triangle whose vertices are $(4, 5)$, $(4, -5)$, $(-4, 5)$. *Ans.* 10; $2\sqrt{41}$; 8.

THE POLAR SYSTEM.

8. Position of a point in a plane. The position of a point on the earth's surface is often designated by its distance and direction from some other point; as when A is said to be 25 miles northeast of B. In a similar way the position of a point in a plane may be designated. Thus: if OA be any assumed straight line through a fixed point O, the position of a point P' in the plane AOP', with reference to O, will be known when the angle AOP' and the distance OP' are known. The fixed line OA is called the **Polar Axis**; the fixed point, O, the **Pole**; the angle AOP' and distance OP', the **Polar Coordinates**, OP' being the **radius vector** and AOP' the **vectorial angle**. The radius vector will always be represented by the letter r, and the vectorial angle by the letter θ.

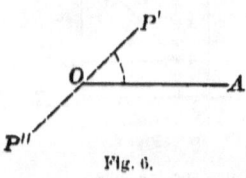

Fig. 6.

THE POINT.   7

9. Signs of the polar coordinates. If the vectorial angle be always laid off above OA (Fig. 6) to the left, as in trigonometry, the position of every point in the plane may be designated without ambiguity, and were this the only consideration there would be no necessity for any convention as to signs. But as r and θ often occur in the course of analytic investigations with negative as well as positive signs, it is necessary to adopt some convention for the interpretation of the negative sign. For this purpose the vectorial angle is regarded positive when laid off above OA to the left, and negative when laid off below OA to the right; while the radius vector is considered positive when laid off from O towards the end of the arc measuring the vectorial angle, and negative when laid off in the opposite direction. Thus (Fig. 6), for the angle AOP', OP' is the positive, and OP'' the negative, direction of r.

10. Construction of a point. Since the coordinates r and θ, when given, fix the position of a point, to determine its position we have only to lay off the given value of θ above or below OA (Fig. 6) according as θ is positive or negative, and on the line through O and the end of the measuring arc the given value of r, towards or away from the end of the measuring arc as the sign of r is positive or negative. This determination of the position of a point is called *the construction of the point*.

11. Equations of a point. It is now evident that we may designate the position of a point by giving its coordinates in the form of equations. Thus (Fig. 6), $r = 4$, $\theta = 60°$, locate a point P' distant 4 units from O on a line inclined at $+60°$ to OA; while $r = -4$, $\theta = 60°$, locate a point P'' on the same line, but on the opposite side of O. These equations are called the *polar equations of a point;* but it is more usual to adopt the notation (r, θ), writing the radius vector first. Thus, the above points would be $(4, 60°)$ and $(-4, 60°)$, respectively.

12. This system of reference is called the **Polar System**, and that previously described, whether the axes be oblique or rectangular, the **Rectilinear System**. It will be observed that in each system *two* things serve as the bases of reference; in the rectilinear, the axes of X and Y; in the polar, the pole and polar axis. Also that in each system *two* quantities are sufficient to refer the point; in the rectilinear system, the abscissa and ordinate; in the polar, the radius vector and vectorial angle. Again, that while in the rectilinear system a given point can have but *one set* of coordinates, in the polar system it may have an *infinite* number of sets. Thus (Fig. 6), P' may be designated as follows: $(4, 60°)$, $(-4, -120°)$, $(-4, 240°)$, $(4, -300°)$, $(4, 420°)$, etc. This fact, however, gives rise to no ambiguity in the position of P', for no one set of polar coordinates can locate more than one point.

The rectilinear and polar systems of reference, together with a third called the *trilinear*, are those in most common use. The two former only will be employed in this treatise.

EXAMPLES. 1. Construct the following points: $(5, 90°)$; $(5, 270°)$; $(-3, 120°)$; $(-6, -180°)$.

2. What are the coordinates of the pole? What are all possible values of θ for points on the polar axis?

3. Construct the points $(0, 45°)$; $\left(0, \dfrac{0°}{0}\right)$; $(4, 0°)$; $(-4, 0°)$.

4. Give three sets of polar coordinates locating $(10, 90°)$.

5. Construct $\left(8, \dfrac{\pi}{4}\right)$; $\left(-8, \dfrac{3\pi}{4}\right)$; $\left(8, \dfrac{9\pi}{4}\right)$; $\left(8, -\dfrac{7\pi}{4}\right)$.

6. The side of a square is $5\sqrt{2}$, its centre at the pole, and sides parallel and perpendicular to the polar axis. What are the coordinates of its vertices?

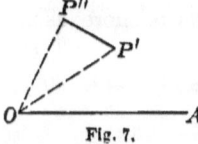
Fig. 7.

13. *To find the distance between two given points.* Let P', P'', be the given points, r', θ', and r'', θ'', their coordinates, and d the required distance. Then, from the triangle $P'OP''$,

$$P'P'' = \sqrt{OP'^2 + OP''^2 - 2\,OP'\cdot OP''\cos P'OP''},$$
or
$$d = \sqrt{r'^2 + r''^2 - 2r'r''\cos(\theta'' - \theta')}. \tag{1}$$

If one of the points, as P'', is at the origin, $d = OP' = r'$.

EXAMPLES. 1. Find the distance between the points $\left(3, \dfrac{\pi}{3}\right)$ and $\left(4, \dfrac{2\pi}{3}\right)$.

Since $\cos\theta = \cos(-\theta)$, it is immaterial which of the two given points is designated as (r', θ'). Thus
$$d = \sqrt{9 + 16 - 24\cos 60°} = \sqrt{16 + 9 - 24\cos(-60°)} = \sqrt{13}.$$
Observe, also, that if $\theta'' - \theta' > 90°$, the cosine will be negative and the last term positive, as it should be, for then the triangle will be obtuse-angled at O (Fig. 7).

2. Find the distances between the following points: $(3, 60°)$ and $(4, 150°)$; $(5, 0°)$ and $(5, -180°)$; $\left(\sqrt{2}, \dfrac{\pi}{4}\right)$ and $(1, 0°)$; $(10, 30°)$ and $(-10, -150°)$; $(6, 60°)$ and $(0, 0°)$

Ans. 5; 10; 1; 0; 6.

SECTION II.—THE LINE.

THE RECTILINEAR SYSTEM.

14. Loci and their equations. *Every line, straight or curved, may be regarded as generated by the motion of a point.* The kind of line generated will depend upon the law which governs the motion of the generating point. Thus, a circle may be traced by a moving point, the law which governs its motion being that it shall always remain at a given distance (the radius) from a fixed point (the centre). If the origin be taken at the centre of the circle, P being *any* point of the circle, x, y, the coordinates of P, and $OP = R$, the radius, then

$$OP^2 = Om^2 + mP^2,$$

or

$$x^2 + y^2 = R^2,$$

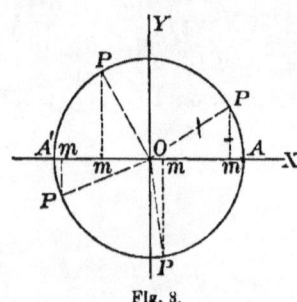

Fig. 8.

is true for every position of P while generating the circle. This equation is the algebraic expression of the law which governs P's motion, and is called the **equation of the circle**; and, in general, *the equation of a line is the algebraic expression of the law which governs the motion of its generating point.*

Again, the relation $x^2 + y^2 = R^2$ is true for no point within or without the circle, but is true for every point on the circle; it thus expresses the relation between the coordinates of all points of the circle, and of no other points; hence, in general, *the equation of a line is the algebraic expression of the relation which exists between the coordinates of any and every point of the line.*

Evidently if a point moves at random, without any governing law, the line it traces can have no equation; for the latter is

THE LINE.

the algebraic expression of a law, and when the point moves at random, none such exists.

The above equation, $x^2 + y^2 = R^2$, being the equation of a circle whose radius is R, the circle is said to be the **locus of the equation**; *i.e.*, translating the word locus literally, it is the *place* in which the point, moving under the law expressed by the equation, is always found; and, in general, *the locus of an equation is the path of a point so moving that its coordinates always fulfil the relation expressed by the equation.*

It follows that if a point lies on a locus, the coordinates of the point must satisfy the equation of the locus. Thus, if the radius of the above circle be 5, $x^2 + y^2 = 25$, and the points $(3, 4)$, $(0, -5)$, $(-4, -3)$, are all points on the circle because their coordinates satisfy the equation; but $(2, 4)$, $(-4, 4)$, are *not* on the circle. Hence, *to ascertain whether a given point lies on a given locus, substitute its coordinates in the equation of the locus and see whether they satisfy it.*

15. Distinctions between Analytic Geometry, Geometry, and Algebra. *The object of Analytic Geometry is the discussion and determination of the properties of loci.* Its method consists in the substitution of the equation of the locus for the locus itself in the discussion and determination of its properties. Thus, $x^2 + y^2 = R^2$ has been seen to be the equation of the circle of Fig. 8. Putting it under the form

$$y^2 = R^2 - x^2 = (R + x)(R - x),$$

we observe that

$$y^2 = Pm^2, \ R + x = A'm, \ R - x = mA.$$

Hence $Pm^2 = A'm \cdot mA$, or the square of the half-chord to any diameter of a circle is a mean proportional between the segments into which it divides that diameter. This well-known property of the circle might be established *geometrically*, from a figure; it is here established *analytically*, from the equation of the circle; and the object of Analytic Geometry is thus to determine the properties of lines, *by discussing their equations,* instead of by reasoning upon the lines themselves as in Euclidean Geometry.

Having thus noted the distinction between Analytic Geometry and Geometry, let us note in what way it differs from Algebra. Since the coordinates of every point of the circle must satisfy its equation $x^2 + y^2 = R^2$, x and y in this equation may have an infinite number of sets of values, corresponding to the infinite number of positions occupied by the generating point in tracing the circle. Hence while R is a *constant* quantity, x and y are *variable* quantities. They differ thus from all the quantities of common Algebra, which, whether known or unknown, are always constants. Observe, also, that while x and y thus admit of an infinite number of values, they do not admit of *any* values, but only of those which satisfy the relation $x^2 + y^2 = R^2$. Again, in Algebra, if only $x^2 + y^2 = R^2$ were given, x and y being unknown but constant quantities whose values were required, the solution would be impossible; for this equation would be satisfied by an infinite number of sets of values of x and y, and without a second independent equation we could not determine the particular values required. Furthermore, if the conditions of the problem were not such as to furnish a second equation, the problem would remain an indeterminate one. It is in virtue of this very indetermination that we are enabled to represent loci by equations, and, as thus distinguished from Algebra, Analytic Geometry is sometimes called the *Indeterminate Analysis*.

16. Quantities of Analytic Geometry. If the centre of the circle were at some point C, whose coordinates are m and n, instead of at the origin, then, from the right-angled triangle PCR, $CR^2 + RP^2 = CP^2$, or
$$(x-m)^2 + (y-n)^2 = R^2.$$

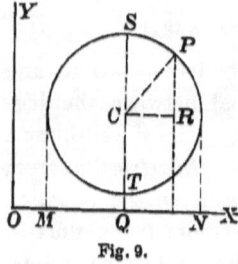

Fig. 9.

This relation between x and y, being true for all positions of P on the circle, is the equation of the circle in its new

THE LINE.

position with reference to the axes. Now if, in this equation of the circle, we change R, we change the *magnitude* of the circle; and if we change m, or n, or both, we change the *position* of the circle. Hence *the constants in the equation of a locus determine the magnitude and position of the locus*. The quantities of Analytic Geometry are thus:

FIRST. *Variable quantities*, as x and y of the preceding equation, which, being the coordinates of a moving point, vary *continuously* within the limits assigned by the equation expressing their mutual relation. Thus x varies continuously between the limits $x = OM$, and $x = ON$, and y between the limits $y = QS$, and $y = QT$. Since, when y changes, x also changes, and *vice versa*, x and y are said to be **functions** of each other.

SECOND. *Arbitrary constants*, as m, n, R, of the above equation, to which values may be assigned at pleasure, thus locating any circle in any position. They do not, however, change *when x and y change*, that is, they are not functions of x and y, and are thus constants, though arbitrary constants.

THIRD. *Absolute constants*, such as m, n, and R, would become in the above equation, if we placed the centre of the circle at (7, 6) and assumed 5 for its radius; which cannot change under any circumstances.

17. Construction of loci. It is now evident that two general classes of problems will arise.

FIRST. *Given the law governing the motion of the generating point* (usually given in the form of some property of the locus), *to find the equation of the locus*.

SECOND. *Given the equation of the locus, to determine the locus; i.e., its position, form, and properties.*

These two fundamental problems form the subject matter of Analytic Geometry and will be fully illustrated in the sequel. Their solution involves on the part of the student a thorough comprehension of the relation between a locus and its equation as defined in Art. 14; and to illustrate this relation the following examples of the determination of loci from their equations by points are added.

14   ANALYTIC GEOMETRY.

EXAMPLES. If any value of either variable, assumed at pleasure, be substituted in the equation of a locus, and the value of the other variable be found from the equation, the set of values thus obtained evidently satisfies the equation; they therefore determine a point of its locus. Hence, to determine points of the locus of an equation, assume in succession any number of values for one variable, and find from the equation the corresponding values of the other. Construct the points thus obtained and draw a line through them. This line will be the locus of the equation. This process is called the *construction of the locus*. The variable to which values are *assigned* is called the *independent variable;* the other, whose values are *derived* from the equation, is called the *dependent variable*. It is evident from the nature of the process that either variable may be chosen as the independent variable, and it is usual to assign values to x and derive those of y. In such an equation, however, as $x = y^3 - 2y^2 + 4$. it is more convenient to assign values to y and derive those of x; *i.e.*, to make the variable which is most involved the independent variable. The illustrations which follow are limited to equations of the first and second degree.

1. $y - x - 4 = 0$. Solving the equation for y, we have $y = x + 4$, and taking x for the independent variable, we obtain for

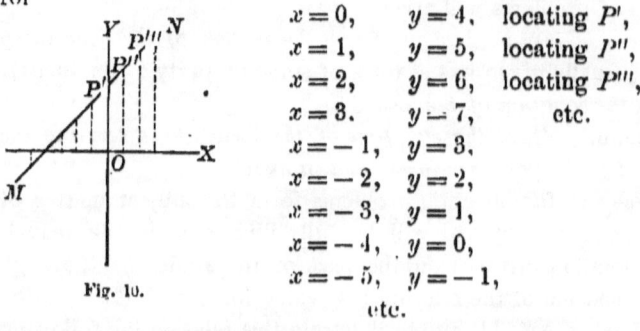

$x = 0$, $y = 4$, locating P',
$x = 1$, $y = 5$, locating P'',
$x = 2$, $y = 6$, locating P''',
$x = 3$. $y = 7$, etc.
$x = -1$, $y = 3$,
$x = -2$, $y = 2$,
$x = -3$, $y = 1$,
$x = -4$, $y = 0$,
$x = -5$, $y = -1$,
etc.

Fig. 10.

Constructing these points, the line MN drawn through them is the locus of $y - x - 4 = 0$.

It will subsequently be shown that the locus of every equation of the first degree, between x and y, is a straight line. This being the case, it is necessary to construct but two points for such equations. Assuming this fact, the student may construct the straight lines represented by the next four equations, constructing in each case two points; then verify the construction by locating a third point.

2. $y + x - 1 = 0$. Solving for y, $y = -x + 1$; in which, for

$x = 0$, $y = 1$, locating P',
$x = 5$, $y = -4$, locating P'',
$x = -3$, $y = 4$, locating P'''.

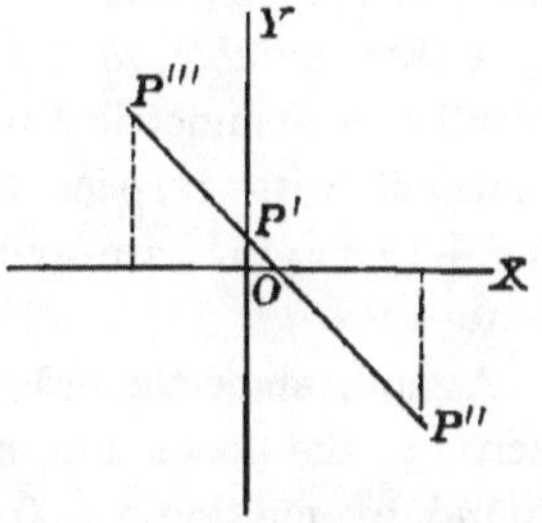

Constructing P', (0, 1), and P'', (5, −4), $P'P''$ should pass through P''', (−3, 4).

3. $y - x = 0$.   4. $y + x = 0$.
5. $3y - 2x - 1 = 0$.

Fig. 11.

6. $y^2 = 4x - 8$. Solving for y, we have, $y = \pm \sqrt{4x - 8}$. Assigning values to x, for $x = 0$ we have $y = \pm \sqrt{-8}$, which is imaginary; moreover y will evidently be imaginary for *all* values of $x < 2$, algebraically. As the ordinates corresponding to all values of $x < 2$ are imaginary, we conclude that there are no points of the locus having abscissas less than 2; and, in general, *when either of the coordinates obtained from the equation is imaginary, we conclude there is no corresponding point of the locus.* Assuming values for $x > 2$, we have, for

$x = 2, y = 0$, locating P^I,
$x = 3, y = \pm 2$, loc. P^II and P^III,
$x = 4, y = \pm 2\sqrt{2}$, loc. P^IV and P^V,
$x = 5, y = \pm 2\sqrt{3}$, etc.
$x = 6, y = \pm 4$,
etc.,

every value of $x > 2$ locating two points.

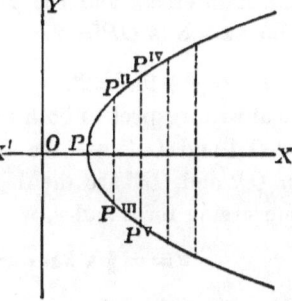

Fig. 12.

The above method of constructing a locus by points is a purely mechanical one. The greater the number of points located, the more accurate the construction of the locus. A simple inspection of the equation will, however, often indicate the general form and position of the locus. Thus, in the above example, every value of x gives two values of y numerically equal with opposite signs, and the locus is therefore made up of pairs of points equidistant from $X'X$, or the axis of X is an *axis of symmetry;* and, in general, *whenever the equation contains the square only of either variable, the other axis is an axis of symmetry.* Thus $y^2 = 9x$ is symmetrical with reference to X; $y + x^2 = 2$ is symmetrical with respect to Y; while $x^2 + y^2 = 25$, $x^2 - y^2 = 4$, $9x^2 + 16x^2 = 144$, are symmetrical with respect to both coordinate axes.

Again: since the ordinate of every point on the axis of X is zero, if the locus has any point on the axis of X it will be found by making $y = 0$; and for a like reason if it has any point on the axis of Y, it will be found by making $x = 0$; and, in general, *to find where a locus crosses or touches either axis, make the other variable zero in its equation.* Thus, in the above example, to find where the locus of $y^2 = 4x - 8$ crosses X, make $y = 0$, whence $x = 2 = OP^1$. Making $x = 0$, y is imaginary, showing that the locus does not meet the axis of Y. *The distances from the origin to the points where a locus meets the axes are called the intercepts of the locus.* They are distinguished as the X-intercept and the Y-intercept. Thus, the X-intercept of $y^2 = 4x - 8$ is $OP^1 = 2$.

7. $25y^2 + 9x^2 = 225$. We observe that the locus is symmetrical with respect to both axes. Making $y = 0$, we find $x = \pm 5$, or OA and OA' are the X-intercepts; making $x = 0$, $y = \pm 3$, or OB and OB' are the Y-intercepts. Solving the equation in succession for x and y, we have

$$y = \pm \tfrac{3}{5}\sqrt{25 - x^2}, \qquad x = \pm \tfrac{5}{3}\sqrt{9 - y^2}.$$

From the value of y we see that x cannot be numerically greater than ± 5, otherwise y is imaginary; hence no point of the curve lies to the right of A or to the left of A'; that is, $x = \pm 5$ gives the *limits* of the curve in the direction of X, and these values are *the roots of the equation obtained by putting the quantity under the radical sign equal to zero.* The reason for this is plain: y is real when $25 - x^2$ is positive, and imaginary when $25 - x^2$ is negative; hence the limiting values of y correspond to $25 - x^2 = 0$, since in passing through zero $25 - x^2$ changes sign. For a like reason, placing $9 - y^2 = 0$, $y = \pm 3$ are the limits of the curve in the direction of Y. And, in general, *whenever the equation of the locus is of the second degree with respect to one of the variables, if we solve it for that variable, and place the radical equal to zero, the roots of this equation are*

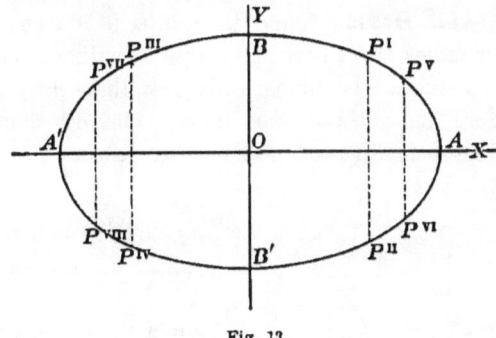

Fig. 13.

the limits in the direction of the other axis. (Thus, in Example 6, the equation is of the second degree with respect to y; solved for y, the radical placed equal to zero gives $4x - 8 = 0$, or $x = 2$. Beyond this limit the curve extends indefinitely in the direction of X.) We have now determined the intercepts, symmetry, and limits, of the locus, and so have a general knowledge of its form and position. Points may now be constructed as before. Thus, for

$x = 3$, or -3, $y = \pm \frac{12}{5}$, locating P^{I}, P^{II}, P^{III}, and P^{IV},

$x = 4$, or -4, $y = \pm \frac{9}{5}$, locating P^{V}, P^{VI}, P^{VII}, P^{VIII}, etc.

8. $16y^2 - 9x^2 = -144$. Making $x=0$, y is imaginary; hence the locus does not meet the axis of Y. Making $y=0$, $x = \pm 4$, or OA and OA', the X-intercepts. The curve is symmetrical with respect to both axes. Solving for x,

$$x = \pm \tfrac{4}{3} \sqrt{y^2 + 9};$$

but $y^2 + 9$ cannot change sign, or, otherwise, $y^2 + 9 = 0$ gives imaginary values for y, hence there are no limits in the direction of Y, the curve extending indefinitely in that direction. Solving for y,

$$y = \pm \tfrac{3}{4} \sqrt{x^2 - 16}.$$

Placing $x^2 - 16 = 0$, the limits in the direction of X are seen to be $+4$ and -4. Having found the limits, it is always necessary to see whether the locus lies within or without the limits. In this case x cannot be numerically less than ± 4, and the curve therefore lies without the limits. Having thus determined the general features of the locus, we proceed to construct a few points. For

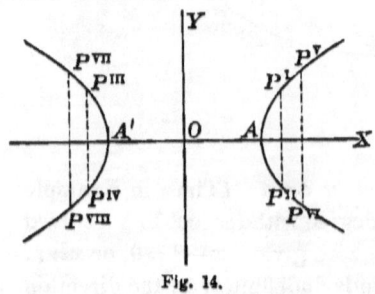

Fig. 14.

$x = \pm 5$, $y = \pm \tfrac{9}{4}$,

locating P^I, P^II, P^III, P^IV,

$x = \pm 6$, $y = \pm \tfrac{3}{2}\sqrt{5}$,

locating P^V, P^VI, P^VII, P^VIII, etc.

A curve of this kind, composed of two separate branches, is said to be *discontinuous*.

9. $x^2 + y^2 - 8x - 4y - 5 = 0$. Making $x=0$, $y=5$, and -1, giving the intercepts OB, OB'. For $y=0$, $x = 4 \pm \sqrt{21} = 4 \pm 4.6$ nearly, or 8.6 and $-.6$ for the intercepts OA, OA'. Solving for x, we have

$$x^2 - 8x = -y^2 + 4y + 5,$$

whence
$$x = 4 \pm \sqrt{-y^2 + 4y + 21}. \quad . \qquad (1)$$

THE LINE.

Now, every value of y gives two values for x of the form $x = 4 \pm p$, and thus locates two points distant p (the radical) from a line parallel to Y and 4 units from it. Thus, for $y = 6$, $x = 4 \pm 3$, locating P^{I} and P^{II}, each distant 3 units from DD', DD' being parallel to Y and 4 units from it. Solving for y, we have

$$y^2 - 4y = -x^2 + 8x + 5,$$

whence
$$y = 2 \pm \sqrt{-x^2 + 8x + 9}, \qquad (2)$$

from which we see the locus is also symmetrical with respect to CC', parallel to X and 2 units above it; and, in general, *whenever the equation, all its terms being transposed to the first member, is of the form $Ax^2 + Bx + $ etc. with respect to either variable, if the coefficient of the square be made positive unity, then half the coefficient of the first power, with its sign changed, will be the distance from the other axis of a line of symmetry parallel to that axis.* Thus, $x^2 + y^2 - 10y + 4 = 0$ is symmetrical with respect to Y, and also with respect to a line parallel to X and 5 units above it; $x^2 + 2x + y^2 - 9y = 0$ is symmetrical with respect to two lines, one parallel to Y at a distance 1 to its left, the other parallel to X at a distance $\frac{9}{2}$ above it.

To find the limits along X, put the radical in (2) equal to zero, whence $x = 9$ and -1. Values of y are imaginary for $x > 9$ or < -1, and the locus lies within these limits. For the limits along Y, (1) gives $y = 7$ and -3, or no point of the locus lies above $+7$ or below -3. Having now determined the intercepts, limits, and symmetry, we may construct a few points. For

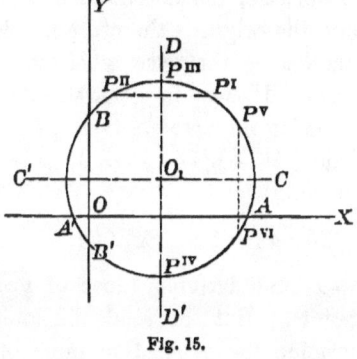

Fig. 15.

$$x = 4, \quad y = 7, \text{ or} - 3, \quad P^{\text{III}}, \text{ and } P^{\text{IV}},$$
$$x = 8, \quad y = 5, \text{ or} - 1, \quad P^{\text{V}}, \text{ and } P^{\text{VI}}, \text{ etc.}$$

20   ANALYTIC GEOMETRY.

10. Show that $y^2 - 6y + x^2 - 16 = 0$, is symmetrical with respect to Y, and a line parallel to X, 3 units above it; that its limits along X are ± 5, and along Y, $+8$ and -2. Determine its intercepts, and construct.

11. $y^2 - 10x + x^2 = 0$. Determine the lines of symmetry, intercepts, limits, and construct.

12. $x^2 - 6x + 9 + y^2 + 10y = 0$. Lines of symmetry are -5 and 3 from X and Y, respectively. Limits along X are 8 and -2; along Y, 0 and -10. Intercepts on Y are -1 and -9; on X, $+3$. Construct.

13. $y^2 - 2x^2 + 12x - 22 = 0$. Show that the locus has no limits in the direction of X, lies wholly outside the limits ± 2 in the direction of Y, has X and a parallel to Y distant $+3$ units from it for lines of symmetry, and $\pm \sqrt{22}$ for Y-intercepts. Construct.

14. $y^2 = 9x$. This locus is symmetrical with respect to X; is without limits along Y, has $x = 0$ for a limit along X, lying wholly in the first and fourth angles. Construct. Observe that if $x = 0$, $y = 0$, and conversely; or the intercepts are zero on both axes, and hence the locus passes through the origin. Otherwise, the coordinates of the origin satisfy the equation, and the origin is therefore a point of the locus. Evidently this cannot be the case when the equation contains an absolute term. Hence, in general, *whenever the equation of the locus contains no absolute term, the locus passes through the origin.* Thus, $x^2 + y^2 - 10y = 0$, $x^4 - y^3 + 3x = 0$ pass through the origin.

15. $xy = 10$. Solving for y, $y = \dfrac{10}{x}$. By assigning values to x, and deriving those of y, we may construct the locus by points. But the student should endeavor in all cases to determine the general features of the locus by an inspection of its equation. In this instance we observe that there is no line of symmetry parallel to either axis, as the equation contains the square of neither variable; also, that y is positive

THE LINE.

when x is positive, and negative when x is negative, and therefore the curve lies wholly in the first and third angles. Again, when $x = 0$, $y = \infty$, and as x increases y diminishes, but becomes zero only when $x = \infty$. In the first angle, then, the locus lies as in the figure, continually approaching the axes as x changes, but touching neither within a finite distance from the origin. *A line to which a curve thus continually approaches, but does not touch within a finite*

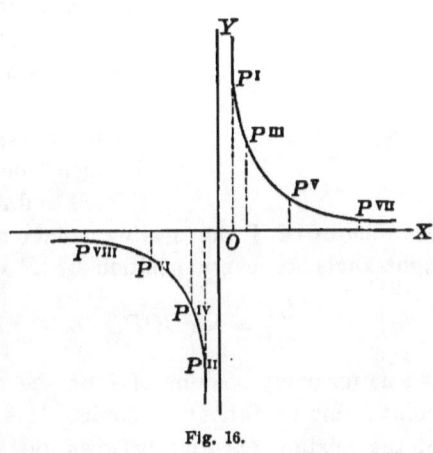

Fig. 16.

distance is called an **asymptote**. In the third angle, x being negative and decreasing algebraically, y increases algebraically, becoming zero, however, only when $x = -\infty$. The axes are thus asymptotes to both branches. Constructing a few points, we have, for

$$x = \pm 1, \quad y = \pm 10, \quad P^\mathrm{I}, P^\mathrm{II},$$
$$x = \pm 2, \quad y = \pm 5, \quad P^\mathrm{III}, P^\mathrm{IV},$$
$$x = \pm 5, \quad y = \pm 2, \quad P^\mathrm{V}, P^\mathrm{VI},$$
$$x = \pm 10, \quad y = \pm 1, \quad P^\mathrm{VII}, P^\mathrm{VIII},$$
$$\text{etc.}$$

THE POLAR SYSTEM.

18. Polar equations of loci. We have seen that the equation of a locus is the algebraic expression of the law governing the motion of the point which traces the locus, and that the quantities in terms of which this law is expressed are the coordinates of the moving point and certain constants. Nothing in this statement restricts us to the use of any particular system of coordinates. Thus, if the law which controls the moving point is that it shall always remain at a given distance from a given point, the line traced will be a circle. C being the fixed point and $CP = R$ the radius or constant distance, if we assume OA as the polar axis, and O the pole at a distance from C equal to the radius, $OP = r$ and $AOP = \theta$ will be the polar coordinates of P the moving point; and since OPB will be a right angle for every position of P while tracing the circle,

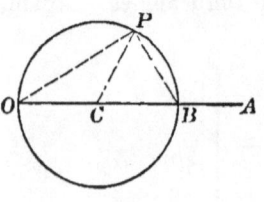

Fig. 17.

$$\frac{OP}{OB} = \cos BOP, \quad \text{or} \quad r = 2R\cos\theta$$

is true for every position of P on the circle, but is true for no point within or without the circle. It is therefore the expression of the relation existing between the coordinates of any and every point of the circle, and is therefore the polar equation of the circle. And, conversely, the circle is the path of a point so moving that its polar coordinates satisfy the above equation; hence the circle is the locus of the equation.

19. Construction of polar equations. In a polar equation, the variables which correspond to x and y of the rectilinear system are r and θ, and by assuming values for one and deriving the corresponding values of the other from the equation, we may construct as many points of the locus as we desire. It is obviously convenient to make θ, the vectorial angle, the independent variable, and derive the values of r.

EXAMPLES. 1. $r = 5$. This equation is independent of θ, that is, $r = 5 = a$ *constant, for all values of* θ. It is then evidently the equation of a circle whose radius is 5, the pole being at the centre. We have seen (Art. 14) that the corresponding rectangular equation of the circle is $x^2 + y^2 = R^2$. The student will observe the comparative simplicity of the polar form $r = R$, and will thus see that in many cases it might be preferable to use the polar rather than the rectangular equation of a locus because of its simpler form.

2. $r = 10\cos\theta$. As in the case of rectangular equations, the student should endeavor to obtain a general idea of the form and position of the locus from its equation, rather than to construct the locus mechanically by points. In the present case we see that when $\theta = 0°$, $\cos\theta$ has its greatest value, and therefore also r; that as θ increases, $\cos\theta$, and therefore also, r, diminishes, becoming zero when $\theta = 90°$. That as θ increases from $90°$ to $180°$, r is negative and increasing numerically, becoming -10 when $\theta = 180°$, the same numerical value which it had for $\theta = 0°$. Constructing a few points, we have, for

$\theta = 0°$,	$r = 10$,	P^{I},
$\theta = 30°$,	$r = 5\sqrt{3}$,	P^{II},
$\theta = 60°$,	$r = 5$,	P^{III},
$\theta = 90°$,	$r = 0$,	O,
$\theta = 120°$,	$r = -5$,	P^{IV},
$\theta = 150°$,	$r = -5\sqrt{3}$,	P^{V},
$\theta = 180°$,	$r = -10$.	

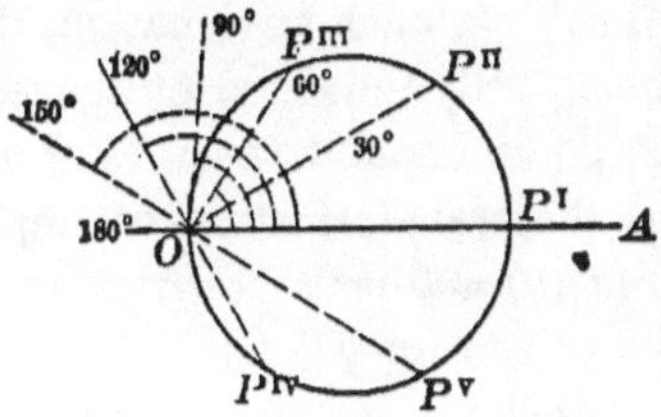

Fig. 13.

As when $\theta = 0°$ the radius vector coincides with the polar axis, P^{I} is constructed by making $OP^{\text{I}} = 10$. Laying off $AOP^{\text{II}} = 30°$, and $OP^{\text{II}} = 5\sqrt{3}$, P^{II} is $(5\sqrt{3}, 30°)$. $\theta = 90°$, gives $r = 0$, and locates the pole. P^{IV} and P^{V} are constructed in the same way, but the values of r when $\theta > 90°$ being negative are laid off away from the end of the measuring arc. If θ increases from $180°$ to $360°$, the values of r are repeated (numerically), so that the entire locus is traced for values of θ from $0°$ to $180°$. As $OP^{\text{I}} = 10$, and $r = 10\cos\theta$ is true for all positions of P, $OP^{\text{II}}P^{\text{I}}$, $OP^{\text{III}}P^{\text{I}}$, etc., is always a right angle, and the locus is therefore a circle whose radius is 10.

The above loci, and those of Art. 17, are constructed simply to familiarize the student with the meaning of the terms *loci of equations*, and, conversely, *equations of loci*. A clear conception of these terms, and of a *coordinate system as a device for representing lines by equations*, is fundamental to the subject. In Chapter II we shall begin the systematic study of loci by means of their equations, commencing with the simplest, namely, the straight line.

20. General notation. Any equation of a locus referred to a rectilinear system of axes may be represented by the equation $f(x, y) = 0$, read 'function x and $y = 0$,' this being a general form for what the equation of the locus becomes when all its terms are transferred to the first member. In such an equation, x and y are said to be **implicit** functions of each other. If the equation of the locus is solved for one of the variables, as y, the corresponding general form will be $y = f(x)$, read 'y a function of x.' In such an equation, the way in which y depends upon x being fully indicated by the solution of the equation, y is said to be an **explicit** function of x. The primary object of Algebra is the transformation of implicit into explicit functions, and $f(x, y) = 0$ may be written $y = f(x)$ whenever the former can be solved for y.

Similarly $f(r, \theta) = 0$, and $r = f(\theta)$, are general forms for the equation of any locus referred to a polar system.

SECTION III.

RELATION BETWEEN THE RECTILINEAR AND POLAR SYSTEMS.

21. Transformation of coordinates. It is evident that the coordinates of a point and the form of the equation of a locus will depend upon the system of reference chosen and its position. Thus, the coordinates of P (Fig. 19) referred to the oblique system $X_1 O_1 Y_1$ are $O_1 m_1$ and $m_1 P$; referred to the rectangular system XOY, they are Om and mP; while if the polar system $O_2 A$ is employed they are $O_2 P$ and $AO_2 P$. Again, we have seen that the equation of a circle referred to rectangular axes through the centre (Fig. 20) is $x^2 + y^2 = R^2$ (Art. 14), but if it is referred to the system $X_1 O_1 Y_1$ its equation is $(x_1 - m)^2 + (y_1 - n)^2 = R^2$

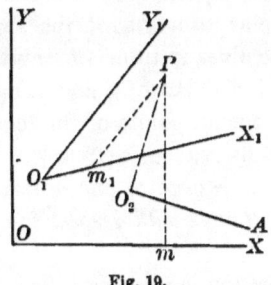
Fig. 19.

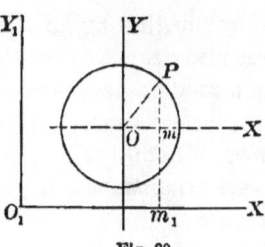
Fig. 20.

(Art. 16), the subscripts being used to distinguish the coordinates of the two systems. Again, in Art. 18, we found the polar equation of a circle to be $r = 2R \cos \theta$ when the pole was on the circle and a diameter was taken for the polar axis; while the polar equation, when the pole was at the centre, was found in Art. 19, Ex. 1, to be $r = R$.

It is thus clear that the form of the equation of any locus will vary with the system of reference chosen, and, from the above illustrations, that one form may be simpler than another. It is therefore desirable to be able to pass from one system to another. This passage from one system of reference to another is called **Transformation of coordinates.**

As this transformation is of frequent use, it is important that the student should thoroughly understand its object and nature. The problem may be thus stated: Having given the equation of a locus referred to one system of reference (as the equation of the circle $(x_1-m)^2+(y_1-n)^2 = R^2$ referred to the axes $X_1 O_1 Y_1$), to find its equation when referred to any other system (as the parallel system XOY, to which when the same circle is referred its equation is $x^2+y^2 = R^2$). The *object* of this transformation is to obtain a simpler equation of the same locus; the *method* will consist in finding values for the coordinates x_1, y_1, in terms of the coordinates x and y, and substituting these values in the given equation; the resulting equation will then be a relation between the new coordinates, and therefore the equation of the locus referred to the new axes.

In the same way, having given the equation of a locus in terms of x and y, we pass to the polar equation of the same locus by substituting for x and y their values in terms of r and θ; the resulting equation will then be independent of x and y, and, being a relation between r and θ true for all points of the locus, is its polar equation. The problem thus reduces to:

The coordinates of any point P with respect to one system of reference being known, to find its coordinates with respect to any other system.

The system to which the transformation is made is called the **new system**; that from which we pass, the **primitive system**. The three following cases will be considered:

(A). To pass from any rectilinear system to any other rectilinear system.

(B). To pass from any rectilinear system to any polar system.

(C). To pass from any polar system to any rectilinear system.

TRANSFORMATION OF AXES.

RECTILINEAR TRANSFORMATIONS.

22. *Formulæ for passing from any rectilinear system to another.*

Let XOY be the primitive system, β being the inclination of the axes, and P any point whose primitive coordinates are $Om = x$, $mP = y$. Let $X_1O_1Y_1$ be the new system, its position being given by the coordinates of its origin, $OA = x_0$, $AO_1 = y_0$, and the angles γ, γ_1, which its axes make with the primitive axis of X, the coordinates of P referred to the new system being $O_1m_1 = x_1$ and $m_1P = y_1$. Draw O_1B and m_1C parallel to OX, and m_1D parallel to OY. Then

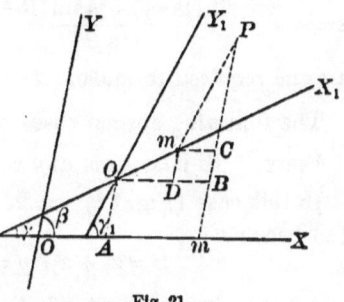

Fig. 21.

$$Om = OA + O_1D + m_1C.$$

But $Om = x$, $OA = x_0$, $O_1D : O_1m_1 :: \sin O_1m_1D : \sin O_1Dm_1$, whence

$$O_1D = \frac{O_1m_1 \sin O_1m_1D}{\sin O_1Dm_1} = \frac{x_1 \sin (\beta - \gamma)}{\sin \beta},$$

and $\qquad m_1C : m_1P :: \sin m_1PC : \sin m_1CP,$

whence $\qquad m_1C = \dfrac{y_1 \sin (\beta - \gamma_1)}{\sin \beta}.$

Substituting these values,

$$x = x_0 + \frac{x_1 \sin (\beta - \gamma) + y_1 \sin (\beta - \gamma_1)}{\sin \beta}.$$

Again, $\qquad mP = AO_1 + Dm_1 + CP.$

But $mP = y$, $AO_1 = y_0$, $Dm_1 : O_1m_1 :: \sin DO_1m_1 \sin O_1Dm_1$,

whence $\qquad Dm_1 = \dfrac{x_1 \sin \gamma}{\sin \beta},$

28   ANALYTIC GEOMETRY.

and $\qquad CP : m_1P :: \sin Cm_1P : \sin m_1CP,$

whence $\qquad CP = \dfrac{y_1 \sin \gamma_1}{\sin \beta}.$

Substituting these values,
$$y = y_0 + \dfrac{x_1 \sin \gamma + y_1 \sin \gamma_1}{\sin \beta}.$$

Hence
$$x = x_0 + \dfrac{x_1 \sin(\beta-\gamma) + y_1 \sin(\beta-\gamma_1)}{\sin \beta},\ y = y_0 + \dfrac{x_1 \sin \gamma + y_1 \sin \gamma_1}{\sin \beta} \quad (1)$$
are the required formulæ.

The following special cases may arise:

FIRST. *To pass from any system to a parallel one.*

In this case (Fig. 22) $\gamma = 0$, $\gamma_1 = \beta$, and the general formulæ (1) become
$$x = x_0 + x_1,\ y = y_0 + y_1, \qquad (2)$$
which are independent of β and apply to all parallel axes, oblique or rectangular.

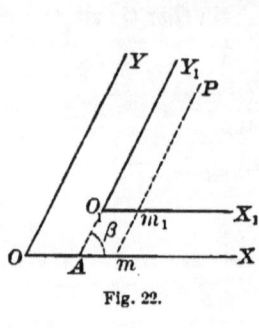

Fig. 22.

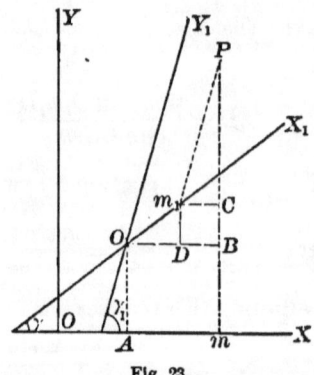

Fig. 23.

SECOND. *To pass from rectangular to oblique axes.*

In this case (Fig. 23) $\beta = 90°$; and since $\sin(90°-A) = \cos A$, $\sin(90°-\gamma)$, and $\sin(90°-\gamma_1)$, become $\cos \gamma$ and $\cos \gamma_1$, or the general formulæ become
$$x = x_0 + x_1 \cos \gamma + y_1 \cos \gamma_1,\quad y = y_0 + x_1 \sin \gamma + y_1 \sin \gamma_1. \quad (3)$$

TRANSFORMATION OF AXES.

THIRD. *To pass from one rectangular system to another.*

In this case (Fig. 24) $\beta = 90°$, $\gamma_1 = 90° + \gamma$; and since

$$\sin(90° + A) = \cos A,$$
$$\sin \gamma_1 = \sin(90° + \gamma) = \cos \gamma,$$
$$\sin(\beta - \gamma) = \cos \gamma,$$
$$\sin(\beta - \gamma_1) = \sin(90° - 90° - \gamma) = \sin -\gamma = -\sin \gamma,$$

and the general formulæ become

$$x = x_0 + x_1 \cos \gamma - y_1 \sin \gamma, \qquad y = y_0 + x_1 \sin \gamma + y_1 \cos \gamma. \qquad (4)$$

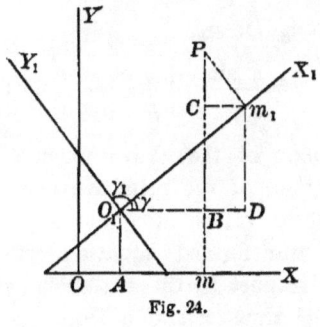
Fig. 24.
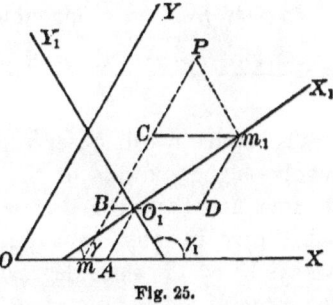
Fig. 25.

FOURTH. *To pass from oblique to rectangular axes.*

In this case (Fig. 25) $\gamma_1 = 90° + \gamma$; and hence

$$\sin(\beta - \gamma_1) = \sin[\beta - (90° + \gamma)] = \sin -[90° - (\beta - \gamma)]$$
$$= -\sin[90° - (\beta - \gamma)] = -\cos(\beta - \gamma),$$

and the general formulæ become

$$\left. \begin{array}{c} x = x_0 + \dfrac{x_1 \sin(\beta - \gamma) - y_1 \cos(\beta - \gamma)}{\sin \beta}, \\ y = y_0 + \dfrac{x_1 \sin \gamma + y_1 \cos \gamma}{\sin \beta}. \end{array} \right\} \qquad (5)$$

The student will observe that the special formulæ, like the general ones, may be deduced directly from the accompanying figures.

If the new origin coincides with the primitive origin, x_0 and y_0 in the above formulæ become zero. Hence,

To pass from one oblique set to another,

$$x = \frac{x_1 \sin(\beta - \gamma) + y_1 \sin(\beta - \gamma_1)}{\sin \beta}, \quad y = \frac{x_1 \sin \gamma + y_1 \sin \gamma_1}{\sin \beta}. \quad (6)$$

To pass from a rectangular to an oblique set,

$$x = x_1 \cos \gamma + y_1 \cos \gamma_1, \quad y = x_1 \sin \gamma + y_1 \sin \gamma_1. \quad (7)$$

To pass from one rectangular set to another,

$$x = x_1 \cos \gamma - y_1 \sin \gamma, \quad y = x_1 \sin \gamma + y_1 \cos \gamma. \quad (8)$$

To pass from an oblique to a rectangular set,

$$x = \frac{x_1 \sin(\beta - \gamma) - y_1 \cos(\beta - \gamma)}{\sin \beta}, \quad y = \frac{x_1 \sin \gamma + y_1 \cos \gamma}{\sin \beta}. \quad (9)$$

The student will observe that none of the above formulæ involve higher powers of the new than of the primitive coordinates, and therefore that when these values of x and y are substituted in any equation, the transformed equation will always be of the same degree with respect to the variables as the primitive equation; that is, the transformation from one rectilinear system to another affects the *form* but not the *degree* of the equation.

POLAR TRANSFORMATIONS.

23. *Formulæ for passing from any rectilinear system to any polar system.* Should the primitive system be oblique, we may first pass to a rectangular system with the same origin by equations (9) of Art. 22; the problem then consists in passing from any rectangular to any polar system.

Let XOY be the primitive system, and P any point whose coordinates are $x = Om$, $y = mP$. Let O_1 be the pole, its coordinates being $OA = x_0$, $AO_1 = y_0$, and let the polar axis make an angle a with the primitive axis of X. Then $O_1P = r$, and

RELATION RECTILINEAR AND POLAR SYSTEMS. 31

$\theta = A_2O_1P$, or A_1O_1P, according as the polar axis lies above or below O_1X_1, drawn parallel to OX, *i.e.* according as a is positive or negative. Hence, in general, $X_1O_1P = \theta \pm a$. Now $Om = OA + O_1D$, in which

$Om = x$, $OA = x_0$,

$O_1D = O_1P \cos DO_1P = r \cos(\theta \pm a)$.

Hence $x = x_0 + r\cos(\theta \pm a)$;
similarly,
$y = y_0 + r\sin(\theta \pm a)$. } (1)

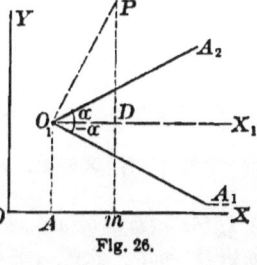
Fig. 26.

If the polar axis is parallel to the axis of X, $a = 0$, *and the general formulæ become*

$$x = x_0 + r\cos\theta, \qquad y = y_0 + r\sin\theta. \qquad (2)$$

If the pole coincides with the origin, $x_0 = y_0 = 0$, *and*

$$x = r\cos(\theta \pm a), \qquad y = r\sin(\theta \pm a). \qquad (3)$$

If the pole is at the origin and the polar axis coincident with X, $a = 0$, $x_0 = y_0 = 0$, *and*

$$x = r\cos\theta, \qquad y = r\sin\theta. \qquad (4)$$

24. *Formulæ for passing from any polar system to any rectilinear system.*

From Equations (1) of Art. 23,

$$x - x_0 = r\cos(\theta \pm a), \qquad y - y_0 = r\sin(\theta \pm a).$$

Squaring, and adding, and substituting the resulting value of r, we have, since $\sin^2 A + \cos^2 A = 1$,

$$\left.\begin{array}{c} r = \sqrt{(x - x_0)^2 + (y - y_0)^2}, \\ \cos(\theta \pm a) = \dfrac{x - x_0}{\sqrt{(x - x_0)^2 + (y - y_0)^2}}, \\ \sin(\theta \pm a) = \dfrac{y - y_0}{\sqrt{(x - x_0)^2 + (y - y_0)^2}}. \end{array}\right\} \quad (1)$$

If the polar axis is parallel to X, $a = 0$, and

$$r = \sqrt{(x-x_0)^2 + (y-y_0)^2}, \quad \cos\theta = \frac{x-x_0}{\sqrt{(x-x_0)^2 + (y-y_0)^2}}, \\ \sin\theta = \frac{y-y_0}{\sqrt{(x-x_0)^2 + (y-y_0)^2}}. \qquad (2)$$

If the new origin is at the pole, $x_0 = y_0 = 0$, and

$$r = \sqrt{x^2 + y^2}, \quad \cos(\theta \pm a) = \frac{x}{\sqrt{x^2+y^2}}, \quad \sin(\theta \pm a) = \frac{y}{\sqrt{x^2+y^2}}. \quad (3)$$

If the new origin is at the pole and the new axis of X coincides with the polar axis, $a = 0$, $x_0 = y_0 = 0$, and

$$r = \sqrt{x^2 + y^2}, \quad \cos\theta = \frac{x}{\sqrt{x^2+y^2}}, \quad \sin\theta = \frac{y}{\sqrt{x^2+y^2}}. \qquad (4)$$

By means of Equations (4) we may pass from any polar system to a rectangular system with the origin at the pole and axis of X coincident with the polar axis; then, by Equations (3) of Art. 22, to any oblique system.

EXAMPLES. 1. Transform $y - x - 4 = 0$ (Ex. 1, Art. 17) to a new set of parallel axes, the new origin being at $(0, 4)$.

The formulæ for passing from any rectilinear system to any parallel one being $x = x_0 + x_1$, $y = y_0 + y_1$, in which $x_0 = 0$, and $y_0 = 4$, the values of the primitive coordinates in terms of the new are, in this case, $x = x_1$, $y = 4 + y_1$. Substituting these values in the given equation, we have $y_1 - x_1 = 0$ for the transformed equation. As the subscripts are only used to distinguish the two sets of coordinates, they may be omitted after the transformation is effected. By referring to Fig. 10, Art. 17, the student will see that the new origin is at P', a point of the locus, and that therefore the transformed equation should have no absolute term.

2. Transform $y + x - 1 = 0$ (Ex. 2, Art. 17) to a new set of parallel axes, the new origin being at $(1, 0)$. *Ans.* $y + x = 0$.

3. Transform $3y - 2x + 4 = 0$ to parallel axes, the new origin being $(-4, -7)$. *Ans.* $3y - 2x - 9 = 0$.

4. Transform $y^2 = 4x - 8$ (Ex. 6, Art. 17) to parallel axes, the new origin being at $(2, 0)$, that is, at P', Fig. 12.
Ans. $y^2 = 4x$.

RELATION RECTILINEAR AND POLAR SYSTEMS. 33

5. Transform $y^2 = 4x - 8$ to a new set of rectangular axes with the same origin, the new axis of X making an angle of $-90°$ with the primitive axis of X.

The formulæ are $x = x_1 \cos \gamma - y_1 \sin \gamma$, $y = x_1 \sin \gamma + y_1 \cos \gamma$, in which $\gamma = -90°$. They become, then, $x = y_1$, $y = -x_1$. Substituting and omitting subscripts, $x^2 = 4y - 8$.

6. Transform $x^2 + y^2 - 8x - 4y - 5 = 0$ (Ex. 9, Art. 17) to a new set of parallel axes, the new origin being at (4, 2), that is, at O_1 (Fig. 15). *Ans.* $x^2 + y^2 = 25$.

7. Transform $xy = 10$ (Ex. 15, Art. 17) to rectangular axes with the same origin, the new axis of X making an angle of $45°$ with the primitive axis of X.

The formulæ are $x = x_1 \cos \gamma - y_1 \sin \gamma$, $y = x_1 \sin \gamma + y_1 \cos \gamma$, in which $\gamma = 45°$, and they become $x = \sqrt{\frac{1}{2}}(x_1 - y_1)$, $y = \sqrt{\frac{1}{2}}(x_1 + y_1)$. Substituting these in $xy = 10$, and omitting subscripts, $x^2 - y^2 = 20$.

8. Transform $x^2 + y^2 = 25$ to a polar system, the pole being at $(-5, 0)$, and the polar axis coincident with X.

The formulæ are $x = x_0 + r \cos \theta$, $y = y_0 + r \sin \theta$, which for $x_0 = -5$, $y_0 = 0$, become $x = -5 + r \cos \theta$, $y = r \sin \theta$. Substituting these in $x^2 + y^2 = 25$, we obtain $r = 10 \cos \theta$.

9. Transform $(x^2 + y^2)^2 = a^2(x^2 - y^2)$ to polar coordinates, the pole being at the origin and the polar axis coincident with X.

Ans. $r^2 = a^2 \cos 2\theta$.

10. Transform the following equations, the origin and the pole being coincident, as also the axis of X and the polar axis. $r = 20 \cos \theta$; $xy = a^2$. *Ans.* $x^2 + y^2 - 20x = 0$; $r^2 = \dfrac{2a^2}{\sin 2\theta}$.

11. Having the distance between two given points in a rectangular system, $d = \sqrt{(x'' - x')^2 + (y'' - y')^2}$ (Art. 7), to find the polar formula for the distance, when the pole is at the origin and the polar axis coincident with X.

Substituting $x' = r' \cos \theta'$, $y' = r' \sin \theta'$, $x'' = r'' \cos \theta''$, $y'' = r'' \sin \theta''$,
$d = \sqrt{(r'' \cos \theta'' - r' \cos \theta')^2 + (r'' \sin \theta'' - r' \sin \theta')^2}$
$= \sqrt{r''^2(\cos^2\theta'' + \sin^2\theta'') + r'^2(\cos^2\theta' + \sin^2\theta') - 2r'r''(\cos\theta''\cos\theta' + \sin\theta''\sin\theta')}$
$= \sqrt{r'^2 + r''^2 - 2r'r'' \cos(\theta'' - \theta')}$,

which is the formula of Art. 13, which was there derived directly from the figure.

12. Under the same conditions find the polar coordinates of the point midway between two given points, having given its rectangular coordinates $\dfrac{x'+x''}{2}$, $\dfrac{y'+y''}{2}$.

Ans. $\dfrac{r'\cos\theta' + r''\cos\theta''}{2}$, $\dfrac{r'\sin\theta' + r''\sin\theta''}{2}$.

13. The distance between two points referred to a rectangular system is $d = \sqrt{(x''-x')^2 + (y''-y')^2}$. Find the distance for an oblique system with the same origin, the new axis of X being coincident with the primitive axis of X, and the new axis of Y making an angle β with it.

The formulæ are $x = x_1 \cos\gamma + y_1 \cos\gamma_1$, $y = x_1 \sin\gamma + y_1 \sin\gamma_1$, which for $\gamma = 0$, $\gamma_1 = \beta$, become $x = x_1 + y_1 \cos\beta$, $y = y_1 \sin\beta$. Substituting these values, and dropping the subscripts,

$$d = \sqrt{(x'' + y''\cos\beta - x' - y'\cos\beta)^2 + (y''\sin\beta - y'\sin\beta)^2}$$
$$= \sqrt{[(x''-x') + (y''-y')\cos\beta]^2 + [(y''-y')\sin\beta]^2}$$
$$= \sqrt{(x''-x')^2 + (y''-y')^2 + 2(x''-x')(y''-y')\cos\beta},$$

a result already obtained in Art. 7.

CHAPTER II.

EQUATION OF THE FIRST DEGREE. THE STRAIGHT LINE.

SECTION IV.—THE RECTILINEAR SYSTEM.

EQUATIONS OF THE STRAIGHT LINE.

25. *Every equation of the first degree between two variables is the equation of a straight line.*

Every such equation may be put under the form
$$Ax + By + C = 0, \qquad (1)$$
in which A and B are the collected coefficients of x and y, and C is the sum of the absolute terms.

Let P', P'', P''', be three points on the locus of this equation, whose abscissas in order of magnitude are x', x'', x'''. Then, from (1), their ordinates y', y'', y''', will also be in order of magnitude. As these three points are on the locus, their coordinates must satisfy its equation; hence

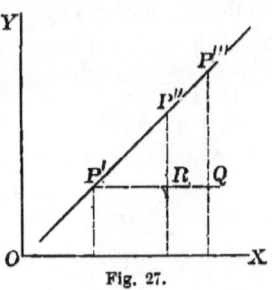

Fig. 27.

$$Ax' + By' + C = 0, \quad Ax'' + By'' + C = 0, \quad Ax''' + By''' + C = 0;$$
whence, by subtraction,
$$A(x'' - x') + B(y'' - y') = 0, \quad A(x''' - x') + B(y''' - y') = 0.$$

Equating the values of A,

$$\frac{y''' - y'}{x''' - x'} = \frac{y'' - y'}{x'' - x'}. \qquad (2)$$

Let $P'Q$ be drawn parallel to OX. Then, from (2),

$$\frac{P'''Q}{P'Q} = \frac{P''R}{P'R}.$$

Hence the triangles $P'''QP'$, $P''RP'$, are similar, and P'' is on the straight line $P'P'''$. In the same manner it may be shown that every other point of the locus is on the same straight line $P'P'''$. The locus is therefore a straight line.

The expression "the line $Ax + By + C = 0$" will frequently be used for brevity, meaning "the line whose equation is $Ax + By + C = 0$."

OBLIQUE AXES. The above demonstration depends only upon the similarity of the triangles and is therefore equally true for an oblique system.

26. Common forms *of the equation of a straight line.* There are three common ways of determining the position of a straight line MN with reference to the axes. *First,* by its intercepts OR, OQ; *second,* by its Y-intercept OQ, and the angle XRQ which the line makes with the axis of X (always measured, as in Trigonometry, from OX to the left); *third,* by the length of the perpendicular OD let fall from the origin on the line, and the angle XOD which this perpendicular makes with the axis of X. In each case the position of the line is evidently completely determined. We are now to find the equation of the line when given in each of these three different ways.

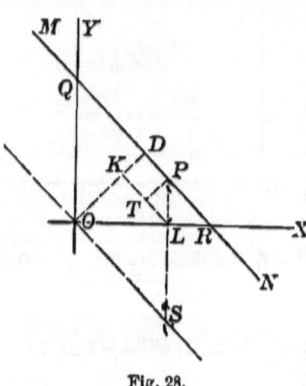

Fig. 28.

FIRST. Let P be *any* point of the line, x, y, its coordinates, and $OR = a$, $OQ = b$, the given intercepts. Then

THE RECTILINEAR SYSTEM.

$$QO:OR::PL:LR, \text{ or } b:a::y:a-x,$$

whence $bx + ay = ab$, or, dividing by ab,

$$\frac{x}{a} + \frac{y}{b} = 1. \qquad (1)$$

SECOND. Draw OS parallel to MN, and let $\tan XRQ = m$. Then $LP = SP - SL$. But

$$SP = OQ = b,$$

$SL = \tan SOL \cdot OL = \tan ORP \cdot OL = -\tan XRQ \cdot OL = -mx$.
Hence $\qquad y = mx + b. \qquad (2)$

The tangent of the angle which the line makes with the axis of X is called the **slope**.

THIRD. Draw LK parallel to MN, and PT parallel to OD. Let $XOD = a$ and $OD = p$. Then $OK + TP = OD$. But
$OK = OL \cos LOK = x \cos a$, $TP = LP \sin TLP = y \sin a$.
Hence $\qquad x \cos a + y \sin a = p. \qquad (3)$

All these equations are, as they should be, of the first **degree** (Art. 25).

Other forms of the equation of a straight line might be found by assuming other constants to fix its position, and such forms will be given later. The reason for employing more than one is that one form is often more convenient than another for the solution of certain problems. Equation (1) is called the **intercept**, Equation (2) the **slope**, and Equation (3) the **normal** form, while the general equation $Ax + By + C = 0$ is called the **general** form.

The student will observe that Art. 25 is an illustration of the general problem: Given the equation, to determine the locus; while this article illustrates the inverse problem: Given the law of the moving point (*straight* line) and the position of the locus (by the constants), to determine its equation. In the latter case, the problem always consists in finding a relation between x and y true for every point of the locus, and expressing this relation in the form of an equation. Whenever we have succeeded in establishing this equation, we have the equation of the locus, whatever the constants involved.

OBLIQUE AXES. Whatever the angle XOY (Fig. 28), the triangles QOR and PLR are similar; hence the intercept form applies without change to oblique axes.

For the slope form we have, as before, $LP = SP - SL$, in which $LP = y$, $SP = b$; but $SL : LO :: \sin SOL : \sin LSO$. Let $\omega = XOY =$ the inclination of the axes, and $\lambda = XRQ$, the angle made by MN with X. Then

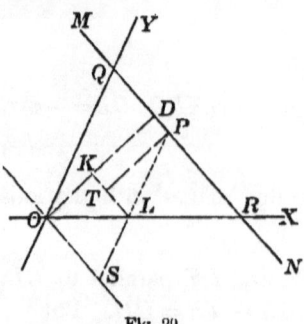

$$SL : x :: \sin \lambda : \sin(\lambda - \omega),$$

whence $SL = -\dfrac{x \sin \lambda}{\sin(\omega - \lambda)}$, and the equation becomes $y = \dfrac{\sin \lambda}{\sin(\omega - \lambda)} x + b$. This may be written in the form $y = mx + b$, understanding that $m = \dfrac{\sin \lambda}{\sin(\omega - \lambda)}$ when the axes are oblique. If $\omega = 90°$, $\sin(\omega - \lambda) = \cos \lambda$, and $m = \tan \lambda$, as in (2).

For the normal form, ROD being α, as before, let $DOQ = \beta$. Then

$$OD = OK + TP = x \cos \alpha + y \cos \beta = p.$$

When $XOY = 90°$, DOQ is the complement of XOD, that is of α, $\cos \beta = \sin \alpha$, and the equation reduces to (3).

Fig. 29.

27. *Derivation of the common forms from the general form.* Since the equation of every straight line is of the general form $Ax + By + C = 0$, it must evidently be possible to derive the common forms from the general form, and to express the particular constants a, b, m, p, $\cos \alpha$, $\sin \alpha$, in terms of the general constants A, B, C.

FIRST. *The intercept form.* Assuming $Ax + By + C = 0$, transposing C to the second member, and dividing the equation by $-C$, *i.e.*, making the second member positive unity, we have

$$\dfrac{x}{\dfrac{-C}{A}} + \dfrac{y}{\dfrac{-C}{B}} = 1,$$

which is the required intercept form, the intercepts being

$$a = -\dfrac{C}{A}, \; b = -\dfrac{C}{B}.$$

SECOND. *The slope form.* Solving the general form for y, we have

$$y = -\dfrac{A}{B} x - \dfrac{C}{B},$$

which is the required slope form, in which $m = -\frac{A}{B}$, $b = -\frac{C}{B}$ as before.

THIRD. *The normal form.* This form requires that the second member (p) should be positive, as no convention has been made for the signs of a distance except as that distance is laid off on the axes; and also that the sum of the squares of the coefficients of x and y should be positive unity, since

$$\cos^2 a + \sin^2 a = 1.$$

Let R be the factor which will transform the general to the normal form. It must fulfil the condition $(RA)^2 + (RB)^2 = 1$. Hence $R = \dfrac{1}{\sqrt{A^2 + B^2}}$. Introducing this factor and transposing C, we have

$$\frac{Ax}{\sqrt{A^2+B^2}} + \frac{By}{\sqrt{A^2+B^2}} = \frac{-C}{\sqrt{A^2+B^2}},$$

in which

$$\cos a = \frac{A}{\sqrt{A^2+B^2}}, \quad \sin a = \frac{B}{\sqrt{A^2+B^2}}, \quad p = \frac{-C}{\sqrt{A^2+B^2}}.$$

To make the second member (p) positive, of the two signs of $\sqrt{A^2+B^2}$ we must evidently take the opposite one of C.

28. In the preceding article we have found the values of a, b, m, p, $\cos a$, and $\sin a$, in terms of A, B, and C. But it is unnecessary for the student to burden his memory with these relations. Thus, suppose we have given the straight line $3x - 4y + 10 = 0$, and its intercepts are required; we have only to put the equation in the intercept form, *i.e.*, transpose the absolute term to the second member and then divide by -10, giving

$$\frac{x}{-\frac{10}{3}} + \frac{y}{\frac{10}{4}} = 1,$$

and the intercepts are seen to be $a = -\frac{10}{3}$, $b = \frac{10}{4}$. A still simpler way of determining the intercepts is to make y and x successively zero (Art. 17, Ex. 6). Thus, for $x = 0$, $y = b = \frac{10}{4}$;

and for $y = 0$, $x = a = -\frac{10}{3}$. Again, suppose the slope is required. We then put the equation under the slope form by solving it for y, obtaining

$$y = \tfrac{3}{4}x + \tfrac{10}{4},$$

and the slope is seen to be $m = \tfrac{3}{4}$, the Y-intercept being $b = \tfrac{10}{4}$, as before. Finally, if the distance of the line from the origin (p) is required, we put the equation under the normal form by dividing it by $\sqrt{A^2 + B^2} = 5$, transposing the absolute term to the second member and changing the signs throughout to make the second member positive, thus obtaining

$$-\tfrac{3}{5}x + \tfrac{4}{5}y = 2,$$

the distance from the origin to the line being 2, and a lying between 90° and 180° since its cosine is negative and sine positive. The exact value of a would be found from the tables, being the angle whose cosine is $-\tfrac{3}{5}$, or sine is $\tfrac{4}{5}$.

29. Discussion of the common forms.

FIRST. *The intercept form.* This form is

$$\frac{x}{a} + \frac{y}{b} = 1, \text{ in which } a = -\frac{C}{A}, \ b = -\frac{C}{B}.$$

If a and b are both positive, the line occupies the position $M^{\text{I}}N^{\text{I}}$ (Fig. 30), both intercepts being laid off in the positive directions of the axes. *If a and b are both negative*, the line occupies the position $M^{\text{II}}N^{\text{II}}$, both intercepts being laid off in the negative directions of the axes. In like manner *when a is positive and b negative*, the line lies as does $M^{\text{III}}N^{\text{III}}$, and *when a is negative and b positive*, as does $M^{\text{IV}}N^{\text{IV}}$. We observe, also, that when C and B, as also C and A, of the general form have like signs, the intercepts are negative, and when they have unlike signs the intercepts are positive. If $a = \infty$, the equation becomes $y = b$, and since y is b for all values of x, that is, since the equation is independent of x, $y = b$ is the equation of a parallel to X at a distance b from it, above or below according as b is positive or negative. Notice that when

$a = \infty$, $A = 0$, and the general form is independent of x. Similarly, *if $b = \infty$*, we have $x = a$, the equation of a parallel to Y, its position to the right or left of Y depending upon the sign of a; in this case $B = 0$, and the general form is independent of y. *If $a = 0$*, the line passes through the origin, therefore b is also zero. In this case the intercept form is inapplicable because there are no intercepts, but we see from the values of a and b that $C = 0$, and the general form becomes $Ax + By = 0$, as it should, since when a locus passes through the origin its equation has no absolute term (Art. 17, Ex. 14).

SECOND. *The slope form.* This form is

$$y = mx + b, \text{ in which } m = -\frac{A}{B}, \; b = -\frac{C}{B}.$$

If m is positive, the line makes an acute angle with X and cuts Y above or below the origin according as *b is positive or negative*. *If m is negative*, the line makes an obtuse angle with X. We thus have

$$y = -mx + b, \quad M^{I}N^{I},$$
$$y = -mx - b, \quad M^{II}N^{II},$$
$$y = mx - b, \quad M^{III}N^{III},$$
$$y = mx + b, \quad M^{IV}N^{IV}.$$

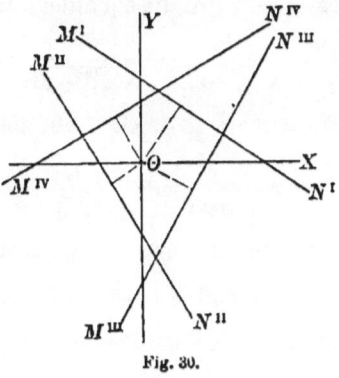

Fig. 30.

If $m = 0$, the line is parallel to X, and $y = b$ is its equation, as already seen. *If $m = \infty$*, the line must be parallel to Y, since the angle whose tangent is ∞ is $90°$. The equation then becomes $y = \infty x + b$. To interpret this form, we observe that, as the line is parallel to Y, b must also be ∞, and hence in $m = -\frac{A}{B}$, $b = -\frac{C}{B}$, the conditions $m = \infty$, $b = \infty$, will both be fulfilled when $B = 0$. The general form then becomes $Ax + C = 0$, or $x = -\frac{C}{A} = a$, as before. *If $b = 0$*, we have $y = mx$, or $Ax + By = 0$, as before, the line passing through the origin. The form $y = mx$ is the most convenient for lines passing through the origin, the value of m fixing the inclination of the line to X.

THIRD. *The normal form.* This form is
$$x \cos a + y \sin a = p,$$
in which
$$\cos a = \frac{A}{\sqrt{A^2 + B^2}}, \quad \sin a = \frac{B}{\sqrt{A^2 + B^2}}, \quad p = -\frac{C}{\sqrt{A^2 + B^2}},$$

the sign of $\sqrt{A^2 + B^2}$ being such as to make p positive. *If both $\cos a$ and $\sin a$ are positive,* a lies between $0°$ and $90°$ ($M^\text{I}N^\text{I}$). *If both are negative,* a lies between $180°$ and $270°$ ($M^\text{II}N^\text{II}$). *If $\cos a$ is positive and $\sin a$ is negative,* a lies between $270°$ and $360°$ ($M^\text{III}N^\text{III}$), and *if $\cos a$ is negative and $\sin a$ positive,* between $90°$ and $180°$ ($M^\text{IV}N^\text{IV}$). If $p = 0$, the line passes through the origin, and its inclination is known when $\sin a$ and $\cos a$ are known, its equation taking the form $x \cos a + y \sin a = 0$,

or
$$\frac{Ax}{\sqrt{A^2 + B^2}} + \frac{By}{\sqrt{A^2 + B^2}} = 0, \text{ or } Ax + By = 0, \text{ as before.}$$

If $a = 0°$ *or* $180°$, $\sin a = 0$, and the equation becomes
$$x = \frac{p}{\cos a} = -\frac{C}{\sqrt{A^2 + B^2}} \cdot \frac{\sqrt{A^2 + B^2}}{A} = -\frac{C}{A} = a,$$

as before, the line being parallel to Y. If $a = 90°$ *or* $270°$, $\cos a = 0$, and $y = \dfrac{p}{\sin a} = b$, in like manner, the line being parallel to X.

OBLIQUE AXES. The intercept form being the same, the discussion above given applies equally to oblique axes.

The slope form is $y = \dfrac{\sin \lambda}{\sin(\omega - \lambda)} x + b$ (Art. 26). Since $\sin \lambda$ is always positive, the sign of the coefficient of x depends upon that of $\sin(\omega - \lambda)$, and will be positive or negative as $\omega >$ or $< \lambda$. Hence $y = \pm \dfrac{\sin \lambda}{\sin(\omega - \lambda)} x \pm b$ represents four lines situated, relative to the axes, as $y = \pm mx \pm b$ in the case of rectangular axes.

The discussion of the normal form $x \cos a + y \cos \beta = p$ (Art. 26) is similar to the above, the equation taking the forms $x = \dfrac{p}{\cos a}$, $y = \dfrac{p}{\cos \beta}$, when the line is parallel to the axes, *i.e.*, when $\beta = 90°$ and $a = 90°$, respectively.

THE RECTILINEAR SYSTEM.   43

30. *To construct a straight line, having given its equation.* If the equation is given in one of the three common forms, we may construct the line by means of the given constants. For example,

$$\frac{x}{3} + \frac{y}{4} = 1, \tag{1}$$

$$y = -\tfrac{4}{3}x + 4, \tag{2}$$

$$\tfrac{4}{5}x + \tfrac{3}{5}y = \tfrac{12}{5}, \tag{3}$$

are the intercept, slope, and normal forms, respectively, of $4x + 3y - 12 = 0$. From the first, make $OR = 3$, $OQ = 4$, and QR is the line. From the second, make $OQ = 4$ and draw QR, making an angle with X, whose tangent is $-\tfrac{4}{3}$. This angle may be constructed without the tables by making $QN = 3$, $NP' = 4$, these lines being parallel to the axes, laying off QN to the right or left of Q, as the angle is acute or obtuse, *i.e.*, as m is plus or minus. From the third lay off $XOD =$ angle whose cosine is $\tfrac{4}{5}$ (or sine $\tfrac{3}{5}$), make $OD = \tfrac{12}{5}$, and draw QR perpendicular to OD.

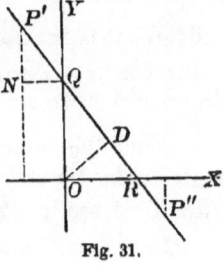

Fig. 31.

Since the line is a straight line, it may evidently be constructed by constructing any two of its points. Thus, for $x = -3$, $y = 8$, (P'), and for $x = 6$, $y = -4$, (P''). But the points most easily constructed are those in which the line crosses the axes. Thus, in any of the above forms, for $x = 0$, $y = 4$, (Q), and for $y = 0$, $x = 3$, (R). Hence, practically, whatever the form in which the equation is given, *to construct a straight line from its equation, construct its intersections with the axes.*

OBLIQUE AXES. To construct the line, make x and y in succession equal to zero, and determine the intercepts.

EXAMPLES. 1. Construct the line whose equation is $x - y = 2$.

Making $y = 0$, we have $x = 2$, the intercept on X; making $x = 0$, we have $y = -2$, the intercept on Y. A line through the points thus found is the required line.

2. Construct the following lines: $y - 2x + 6 = 0$; $x + y = 7$; $3y + x - 9 = 0$; $3x + 5y + 15 = 0$.

3. Construct the line $y + 2x = 0$.

Since this equation has no absolute term the line passes through the origin. In such a case construct any second point; thus $x = 1$ gives $y = -2$. Then join $(1, -2)$ with the origin.

4. Construct the lines $y = x$; $y = -x$; $2y - 3x = 0$.

5. What angle does $y + 3x - 7 = 0$ make with X?

Putting the equation under the slope form, $y = -3x + 7$, and the angle is the angle whose tangent is -3.

6. What is the distance of $6x + 8y + 11 = 0$ from the origin?

Dividing by $-\sqrt{A^2 + B^2} = -10$, we have $-\tfrac{3}{5}x - \tfrac{4}{5}y = \tfrac{11}{10}$, and $p = \tfrac{11}{10} =$ the required distance.

7. Find the intercepts, slope, distance from the origin, and angle made by the perpendicular from the origin on the line with X, of the line $2x + 7y - 9 = 0$.

Making $x = 0$, $y = b = \tfrac{9}{7}$; making $y = 0$, $x = a = \tfrac{9}{2}$. Solving for y, $y = -\tfrac{2}{7}x + \tfrac{9}{7}$, $\therefore m = -\tfrac{2}{7}$. The normal form is

$$\frac{2x}{\sqrt{53}} + \frac{7y}{\sqrt{53}} = \frac{9}{\sqrt{53}}, \therefore p = \frac{9}{\sqrt{53}}, \text{ and } a = \sin^{-1}\frac{7}{\sqrt{53}}.$$

8. Find a, b, m, p, and a in the following lines:

$3y - 4x + 25 = 0$; $7x - y = 0$; $y + x - 3 = 0$.

9. The intercepts of a line are 6, 3; write its equation.

Ans. $x + 2y - 6 = 0$.

10. A line makes an angle of 45° with X, and cuts Y at -2 from the origin; write its equation. *Ans.* $y = x - 2$.

11. The distance of a line from the origin is 6, and the perpendicular upon it from the origin makes an angle of 60° with X; write the equation of the line. *Ans.* $\sqrt{3}y + x = 12$.

12. Same as Ex. 11, when the angle is 120°.

13. Write the equations of parallels to X, one 4 above, and one 10 below it.

THE RECTILINEAR SYSTEM.

14. Write the equations of the sides and diagonals of a square whose side is 10, the sides being parallel to the axes and the centre at the origin.

15. What are the equations of the axes?

Since $x = a$ is a parallel to Y at a distance $= a$, if $a = 0$ the line coincides with Y. Hence $x = 0$ is the equation of Y. Similarly $y = 0$ is the equation of X.

16. Determine which of the points $(2, 3)$, $(1, -3)$, $(-2, 7)$, $(-3, 11)$ are on the line $2y + 7x - 1 = 0$.

17. Find the length of the portion of the line $4x + 3y = 24$ included between the axes. *Ans.* 10.

18. The line $y = mx$ passes through (x', y'); find the value of m.

31. *Equation of a straight line passing through a given point.*

Let (x', y') be the given point. Since the required line is a straight line, its equation will be of the form (1) $Ax + By + C = 0$, and since it passes through the point (x', y'), we have $Ax' + By' + C = 0$. Subtracting this from (1),

$$A(x-x') + B(y-y') = 0, \therefore y-y' = -\frac{A}{B}(x-x'). \text{ But } -\frac{A}{B} = m.$$

Hence $\qquad y - y' = m(x - x'), \qquad\qquad (2)$

being a relation between x and y in terms of the given constants x', y', is the required equation.

An infinite number of lines may be drawn through a given point; hence the line is not determined unless its slope m is also given. Thus, the line through $(1, -4)$, making an angle of $45°$ with X, is $y + 4 = 1(x-1)$, or $y = x - 5$.

OBLIQUE AXES. The above equation applies to oblique axes, understanding that
$$m = \frac{\sin \lambda}{\sin (\omega - \lambda)}.$$

32. *Equation of a straight line passing through two given points.*

Let (x', y'), (x'', y'') be the given points. The required equation will be of the form (1) $Ax + By + C = 0$, since the line is a straight line, and must be satisfied for the co-ordinates of the given points; hence (2) $Ax' + By' + C = 0$, (3) $Ax'' + By'' + C = 0$. Subtracting (2) from (1), and (3) from (2), we have

$$A(x - x') + B(y - y') = 0, \quad A(x' - x'') + B(y' - y'') = 0.$$

Transposing, and dividing,

$$y - y' = \frac{y' - y''}{x' - x''}(x - x'), \tag{4}$$

which is a relation between x and y in terms of the given constants, and hence the equation required, the coefficient of x, $\frac{y' - y''}{x' - x''}$, being the slope (Art. 27). Thus the line passing through (1, 2) and (-3, 4) is $y - 2 = \frac{2 - 4}{1 + 3}(x - 1)$, or $2y + x - 5 = 0$. It is immaterial which point is designated as (x', y'). Thus, $y - 4 = \frac{4 - 2}{-3 - 1}(x + 3)$, or $2y + x - 5 = 0$, as before.

OBLIQUE AXES. Nothing in the above reasoning being dependent upon the inclination of the axes, the equation is the same if the axes are oblique, only the coefficient $\frac{y' - y''}{x' - x''}$ is then the ratio of the sines of the angles which the line makes with X and Y.

EXAMPLES. 1. Write the equation of a line through (2, 4) having the slope 5. *Ans.* $y - 5x + 6 = 0$.

2. Write the equation of a line through (2, 3) and (1, -2).

In place of using Eq. (4), Art. 32, as there illustrated, it is quite as expeditious to determine the constants of any one of the three common forms directly. Thus, the form $y = mx + b$, satisfied in succession for the two points gives

$$3 = 2m + b, \quad \text{and} \quad -2 = m + b.$$

Subtracting, we obtain $m = 5$. Substituting this value of m in either of the above, we find $b = -7$. Hence $y = 5x - 7$.

3. Find the equations of the sides of the triangle whose vertices are (4, 8), (1, 4), (-4, -8).

Ans. $3y - 4x - 8 = 0$; $5y - 12x - 8 = 0$; $y - 2x = 0$.

THE RECTILINEAR SYSTEM. 47

4. Find the equations of the medials of the triangle of Ex. 3.
Ans. $11y - 20x - 8 = 0$; $y - 4x = 0$; $13y - 28x - 8 = 0$.

5. Write the equation of a line through (2, 5) and the origin.

We may use the equation of a line through two points, making one of the points (0, 0); or the slope form $y = mx$ (since $b = 0$), which, satisfied for (2, 5), gives $m = \frac{5}{2}$, and therefore $y = \frac{5}{2}x$.

6. Write the equations of the following lines :
  (1) through $(-7, 1)$, making an angle 45° with X.
  (2) through $(2, -1)$, and $(-3, 4)$.
  (3) through $(-1, -7)$, and the origin.
  (4) through $(-6, -3)$, parallel to X.
  (5) through $(-1, 2)$, parallel to Y.

Ans. $y - x - 8 = 0$; $y + x - 1 = 0$; $y = 7x$; $y = -3$; $x = -1$.

PLANE ANGLES.

33. *To find the angle included between two given straight lines.*

The slope form is best adapted to this problem.

Let $y = mx + b$, $y = m'x + b'$, be the two given lines; m and m' are the tangents of the angles $XRP = \lambda$, and $XQP = \lambda'$. Then, if $c = \tan RPQ = \tan\gamma$, from Trigonometry we have

$$\tan\gamma = \frac{\tan\lambda' - \tan\lambda}{1 + \tan\lambda \tan\lambda'},$$

or $\qquad c = \dfrac{m' - m}{1 + mm'}.$ (1)

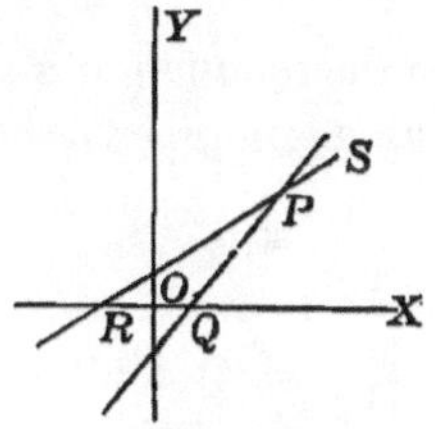

Fig. 32.

Thus the tangent of the angle between $y = 4x + 7$ and $y = 2x - 1$ is $c = \dfrac{4 - 2}{1 + 8} = \dfrac{2}{9}$, and the angle may be found from a table of natural tangents. It is immaterial whether we substitute 4 for m' and 2 for m, or *vice versa*; in the latter case $c = \dfrac{2 - 4}{1 + 8} = -\dfrac{2}{9}$, the difference in sign being due to the fact that the tangents of the supplementary angles RPQ

and QPS which the lines make with each other are numerically equal with opposite signs. We thus obtain the acute or obtuse angle, according as the sign of the result is positive or negative.

34. *To find the equation of a straight line making a given angle with a given line.*

Let $y = mx + b$ be the given line, $y = m'x + b'$ the required line, and $c =$ tangent of the given angle. Then in the relation $c = \dfrac{m' - m}{1 + mm'}$, c and m are known. Solving for m', we have $m' = \dfrac{m + c}{1 - mc}$. Hence the required equation is

$$y = \frac{m + c}{1 - mc} x + b'. \tag{1}$$

Since an infinite number of straight lines may be drawn making a given angle with a given line, b' is undetermined. We are then at liberty to impose another condition upon the line, as that it shall pass through a given point. The equation of a line through a given point is $y - y' = m'(x - x')$, in which m' may have any value (Art. 31). Substituting the value found above,

$$y - y' = \frac{m + c}{1 - mc}(x - x') \tag{2}$$

is the *equation of a straight line passing through a given point and making a given angle with a given line.* Thus, the line through (2, 4), making an angle 45° with $y = 2x - 4$, is $y - 4 = \dfrac{2 + 1}{1 - 2}(x - 2)$, or $y = -3x + 10$.

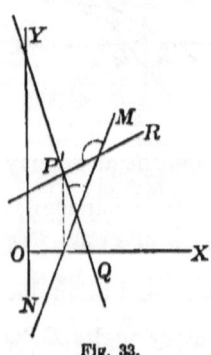

Fig. 33.

Constructing, MN is the given line $y = 2x - 4$; P' the given point (2, 4); and $P'Q$ the line $y = -3x + 10$. The student will observe that $P'R$ makes an angle 135° with MN, the angle being measured, as always, from the line to the left; and that, therefore, to obtain the equation of $P'R$ we should make $c = \tan 135° = -1$.

THE RECTILINEAR SYSTEM. 49

35. *Conditions that two lines shall be parallel, or perpendicular, to each other.*

FIRST. *If two lines are parallel,* their included angle is zero. Hence (Art. 33) $c = \dfrac{m'-m}{1+mm'} = 0$, or $m = m'$; that is, *two lines are parallel whenever, their equations being solved for y, the coefficients of x are equal.* This follows obviously from the fact that parallel lines make equal angles with X, and hence the tangents of these angles, m, m', must be equal. Thus,

$y = 2x + 4,\ y = 2x - 7,\ y - 2x = 0$, are all parallels.

Cor. The equation of a line passing through a given point (x', y') parallel to a given line $y = mx + b$, is (Art. 31)

$$y - y' = m(x - x'). \qquad (1)$$

SECOND. *If the lines are perpendicular* to each other, their included angle is 90°, and hence $c = \dfrac{m'-m}{1+mm'} = \infty$,

or $\qquad 1 + mm' = 0,\ \therefore m' = -\dfrac{1}{m}$;

that is, *two lines are perpendicular to each other whenever, their equations being solved for y, the coefficients of x are negative reciprocals of each other.* Thus,

$y = -\tfrac{2}{3}x + 4,\ y = -\tfrac{2}{3}x - 6,\ y + \tfrac{2}{3}x = 0$,

are all perpendicular to $\quad y = \tfrac{3}{2}x + 7$.

Cor. The equation of a line passing through a given point (x', y') perpendicular to a given line $y = mx + b$

is $\qquad y - y' = -\dfrac{1}{m}(x - x'). \qquad (2)$

The equations $m = m',\ m' = -\dfrac{1}{m}$, are not the equations of lines, for they contain no variables. Since they involve only constants (which serve to fix the position of the lines in question), they express conditions imposed upon the position of the lines. Such equations are called *equations of condition.*

EXAMPLES. 1. Find the angles between the lines $y = -x + 2$, $y = 3x + 7$; $y = \frac{2}{3}x - 1$, $y = \frac{4}{7}x + 4$; $y = 2x - 3$, $y = 2x + 7$; $y = \frac{1}{2}x - 3$, $y = -2x + 9$; $3y + 4x + 1 = 0$, $2y + x + 5 = 0$.

Ans. $c = 2$; $c = \frac{4}{33}$; $0°$; $90°$; $c = \frac{1}{2}$.

2. Write the equation of a line making an angle whose tangent is 3 with $y = -3x + 4$. *Ans.* $y = b$.

3. Write the equations of lines making angles of $45°$ and $135°$ with $2y - x + 3 = 0$. *Ans.* $y = 3x + b$; $y = -\frac{1}{3}x + b$.

4. Write the equation of a line through $(-3, 7)$ making an angle whose tangent is $\sqrt{3}$ with $2y - x + 1 = 0$.

Ans. $(2 - \sqrt{3})y - (1 + 2\sqrt{3})x - 17 + \sqrt{3} = 0$.

5. Write the equations of two parallels to $y = \frac{4}{3}x + 7$; also to $3y + 7x = 0$.

6. Write the equation of a parallel to $3y - 4x = 2$ through $(1, 2)$. *Ans.* $3y - 4x - 2 = 0$.

7. Write the equations of two perpendiculars to
$y = -\frac{1}{2}x + 4$; also to $y - x + 4 = 0$.

8. Write the equation of a line through $(7, -1)$ perpendicular to $y = -4x + 1$; also through $(7, -1)$ perpendicular to $3y - 2x = 0$. *Ans.* $4y - x + 11 = 0$; $2y + 3x - 19 = 0$.

9. Write the equations of lines through $(1, 3)$ making angles of $0°$, $90°$, and $45°$ with X. *Ans.* $y = 3$; $x = 1$; $y - x - 2 = 0$.

10. Write the equation of a line through $(5, 3)$ parallel to the line whose intercepts are 3, 2. *Ans.* $3y + 2x - 19 = 0$.

11. Write the equation of a line through $(2, 3)$ perpendicular to the line joining $(2, 1)$ with $(-2, 5)$. *Ans.* $y = x + 1$.

12. The vertices of a triangle are $(-1, -1), (-3, 5), (7, 11)$. Write the equations of its altitudes.

Ans. $3y - x - 26 = 0$; $3y + 5x + 8 = 0$; $3y + 2x - 9 = 0$.

13. Write the equations of the perpendiculars erected at the middle points of the sides of the triangle of Ex. 12.

Ans. $3y - x - 8 = 0$; $3y + 5x - 34 = 0$; $3y + 2x - 21 = 0$.

14. Prove that $Ax + By + C = 0$ is perpendicular to
$A'x + B'y + C' = 0$ if $AA' + BB' = 0$.

15. Prove that $Ax + By + C = 0$ is parallel to
$A'x + B'y + C' = 0$ if $AB' - A'B = 0$.

16. Prove that the angle between $Ax + By + C = 0$ and $A'x + B'y + C' = 0$ is given by the relation $\tan \gamma = \dfrac{A'B - AB'}{AA' + BB'}$.

17. Write the equation of a straight line perpendicular to $Ax + By + C = 0$ and making an intercept a on the axis of X.

18. Write the equation of a line perpendicular to $y = mx + b$ and at a distance d from the origin.

19. Write the equation of a line parallel to $y = mx + b$ and at a distance d from the origin.

20. Prove that if the equations of two straight lines differ only in their absolute terms, the lines are parallel.

INTERSECTIONS.

36. Intersection of loci. The point of intersection of two straight lines is the point common to both. But if a point lies on a given straight line, its coordinates must satisfy the equation of the line; hence the coordinates of the point of intersection must satisfy the equations of each line. Conversely, *to find the point of intersection of two straight lines, combine their equations and find the set of values of the coordinates which satisfies them both.*

The above reasoning is obviously entirely general. Whatever the loci under consideration, if they have a common point, or points, the coordinates of these points must satisfy both equations. Hence, in general, *to find the intersections of any two loci, combine their equations.*

Since the number of sets of values of x and y, obtained by making the equations simultaneous, is equal to the product of the numbers indicating the degrees of the equations, this product also indicates the possible number of intersections. If, for example, the equations are of the second degree, their loci may intersect in four points, but no more; and as some of the values of x and y thus obtained may be imaginary, the number of *real* intersections may be less than four. And, in general, the greatest possible number of intersections of two loci whose equations are of the pth and qth degrees, respectively, will be pq, and the number of real intersections will be the number of sets of coordinates, satisfying both equations, in which x and y are both real.

Since all equations of straight lines are of the first degree, but one set of values of x and y can be found satisfying any two such equations, or two straight lines can intersect in but one point. If two straight lines are parallel, they cannot intersect, and the combination of their equations will give an impossible result. Thus, $x + y = 4$ and $x + y = 7$ are parallels. Combining, we obtain $3 = 0$. Hence *non-intersection is shown by the occurrence of impossible or imaginary results*. Otherwise, equating the values of x, $0y = 3$, or $y = \frac{3}{0} = \infty$, showing that the lines intersect only at an infinite distance.

EXAMPLES. Find the intersection of the following lines:

1. $2y - 3x - 7 = 0$, and $2y + x - 10 = 0$.   Ans. $(\frac{3}{4}, \frac{37}{8})$.

2. $x + 2y - 5 = 0$, and $2x + y - 7 = 0$.   Ans. $(3, 1)$.

3. $y - x + 1 = 0$, and $y + x + 1 = 0$.   Ans. $(0, -1)$.

4. $6x + 6y - 1 = 0$, and $x + y = 4$.

5. $x + y = 0$, and $x - y = 0$.

6. Find the vertices of the triangle whose sides are
   $5y - 12x - 8 = 0$, $3y - 4x - 8 = 0$, $y - 2x = 0$.
   Ans. $(1, 4)$, $(-4, -8)$, $(4, 8)$.

THE RECTILINEAR SYSTEM.   53

7. Show that $y + 3x - 1 = 0$, $y + 2x + 7 = 0$, $y - x + 31 = 0$, meet in a point.

8. Show that the medials of the triangle of Ex. 4, Art. 32, meet in a point.

9. Show that the altitudes of the triangle of Ex. 12, Art. 35, meet in a point.

10. Show that the perpendiculars erected at the middle points of the sides of the triangle of Ex. 13, Art. 35, meet in a point.

Examples on the intersection of curves are reserved until the student is familiar with the equations of the curves; but he will observe that the process is the same, whatever the degree of the equations or the system of reference: *to find the intersections of any lines, combine their equations.*

37. *Lines through the intersection of loci.*

Let (1) $Ax + By + C = 0$, (2) $A'x + B'y + C' = 0$, be the equations of any two straight lines, and k any arbitrary constant; then is (3) $Ax + By + C + k(A'x + B'y + C') = 0$ the equation of a straight line through their intersection. For the values of x and y which satisfy (1) and (2), evidently satisfy (3) also, hence (3) passes through the point of intersection of (1) and (2). Moreover (3) is of the first degree, hence the equation of a straight line.

NOTE. This reasoning is entirely independent of the form and the degree of the equations. Hence *if* $a = 0$, $\beta = 0$, *be the equations of any two loci,* a *and* β *representing any functions of* x *and* y, *and* k *be any arbitrary constant,*
$$a + k\beta = 0$$
passes through all their points of intersection.

So long as k is arbitrary, (3) will represent *any* straight line through the intersection of (1) and (2), and as k may have any value, it may be determined so that (3) shall fulfil any reasonable condition. Thus:

First. *To find the equation of a straight line passing through the intersection of two given straight lines and also through a given point.* Let (1) and (2) be the two given lines and (x', y') the given point. Then (3) is a straight line through their intersection. Since this line is to pass through (x', y'), we have $Ax' + By' + C + k(A'x' + B'y' + C') = 0$, in which everything is known but k. Determining k from this equation and substituting its value in (3), we have the required equation.

Second. *To find the equation of a straight line passing through the intersection of two given straight lines, and parallel (or perpendicular) to a given line.* Let (1) and (2) be the given lines and $y = mx + b$ the line to which (3) is to be parallel (or perpendicular). Solve (3) for y and place the coefficient of x equal to $m \left(\text{or} -\frac{1}{m}\right)$; from this equation determine k and substitute its value in (3). The resulting equation will be the line required.

EXAMPLES. Write the equations of the following lines:

1. Through the intersection of
$$x + 2y - 5 = 0 \text{ and } y - 3x + 8 = 0,$$
and the point (6, 4).

Substituting $x = 6$, $y = 4$, in $x + 2y - 5 + k(y - 3x + 8) = 0$, we find $k = \frac{9}{6}$. Hence the required line is $y - x + 2 = 0$.

2. Through the intersection of
$$2x + y - 7 = 0 \text{ and } x + 2y - 5 = 0,$$
parallel to $6x - 3y + 5 = 0$.

We have $2x + y - 7 + k(x + 2y - 5) = 0$. Solving for y,
$$y = -\frac{2+k}{1+2k}x + \frac{7+5k}{1+2k}.$$

Solving the parallel for y, $y = 2x + \frac{5}{3}$. Hence $-\frac{2+k}{1+2k} = 2$. $\therefore k = -\frac{4}{5}$, and the required line is $2x - y - 5 = 0$.

3. Through the intersection of
$$2x + y - 7 = 0 \text{ and } y - x - 1 = 0,$$
perpendicular to $3x + 3y - 1 = 0$. *Ans.* $y - x - 1 = 0$.

THE RECTILINEAR SYSTEM.

4. Through the intersection of
$$y - x - 1 = 0 \text{ and } y - 2x + 1 = 0,$$
parallel to $y = 4x + 7$. *Ans.* $y = 4x - 5$.

5. Through the intersection of
$$y = 3x + 14 \text{ and } y = x + 6,$$
making an angle of 45° with $y = 2x$. *Ans.* $y = -3x - 10$.

6. The line $y = mx + b$ passes through the intersection of $y = m'x + b'$ with $y = m''x + b''$. Find the value of m.

DISTANCES BETWEEN POINTS AND LINES, AND ANGLE-BISECTORS.

38. *To find the distance of a given point from a given straight line.*

Let $x \cos a + y \sin a = p$ be the given line and (x', y') the given point. Through the given point, P', draw ST parallel to the given line MN. The perpendiculars OQ, OR, from the origin on these lines, coincide; therefore a is the same for both (Art. 26), and the equations of the parallels will differ only in the lengths of the perpendiculars. Hence, if $OR = p'$, the equation of ST will be

$$x \cos a + y \sin a = p',$$

and since P' is on ST,

$$x' \cos a + y' \sin a = p'.$$

Now $DP' = QR$ is the difference between p' and p, hence the required distance is

$$D = x' \cos a + y' \sin a - p.$$

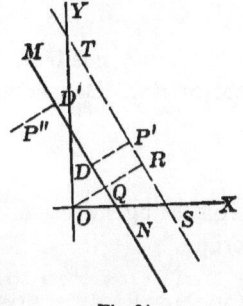

Fig. 34.

But this is simply what the equation of the given line becomes when p is transposed to the first member and x', y', substituted for x, y. Hence, *to find the distance of a given point from a given line, put the equation of the given line under the normal form, transpose the absolute term to the first member, and substitute the coordinates of the given point.* Since to put $Ax + By + C = 0$ under the normal form we divide by $\sqrt{A^2 + B^2}$, we have

$$D = \frac{Ax' + By' + C}{\sqrt{A^2 + B^2}}. \qquad (1)$$

As only the distance DP' is required, it is not *necessary* to attend to the sign of $\sqrt{A^2 + B^2}$. If, however, we follow the general rule of signs for putting the general under the normal form (Art. 27), the last term of (1), $\frac{C}{\sqrt{A^2 + B^2}}$, will always be *negative*, since, when transposed, it must equal $+p$. Now if we make x' and y' zero, (1) will be the distance of the origin from the line $= \frac{C}{\sqrt{A^2 + B^2}}$; hence the origin is always considered as being on the negative side of the line. Whenever, then, for any given point, (1) is negative, the point is on the same side of the line as the origin. Thus, suppose the equation of MN is $2x + y - 2 = 0$. Dividing by $\sqrt{5}$, the normal form is $\frac{2x}{\sqrt{5}} + \frac{y}{\sqrt{5}} - \frac{2}{\sqrt{5}} = 0$. Substituting the coordinates of P', $(\frac{3}{2}, \frac{3}{2})$,

$$D = \frac{2 \cdot \frac{3}{2} + \frac{3}{2} - 2}{\sqrt{5}} = \frac{2\frac{1}{2}}{\sqrt{5}} = DP'.$$

This being positive, P' is on the opposite side of the line from the origin. But, substituting the coordinates of P'', $(-\frac{3}{2}, 2)$,

$$D = \frac{-2 \cdot \frac{3}{2} + 2 - 2}{\sqrt{5}} = -\frac{3}{\sqrt{5}} = P''D'.$$

This being negative, P'' is on the same side of the line as the origin.

<small>Were the axes oblique, the equation of the given line being $x \cos a + y \cos \beta - p = 0$ (Art. 26), as the reasoning above is independent of β, $x' \cos a + y' \cos \beta - p$ would be the required distance.</small>

39. The distance from a given point to a given line may also be found as follows: Write the equation of a line through the

THE RECTILINEAR SYSTEM.

given point perpendicular to the given line; find the intersection of the perpendicular and the given line; then find the distance from the given point to this intersection by the formula $d = \sqrt{(x'-x'')^2 + (y'-y'')^2}$. Thus, to find the distance from $(8, 1)$ to $3x - 4y + 5 = 0$; the perpendicular through $(8, 1)$ to $3x - 4y + 5 = 0$ is $y - 1 = -\frac{4}{3}(x-8)$; combining this with $3x - 4y + 5 = 0$, we have for the point of intersection $(5, 5)$. Hence $d = \sqrt{(8-5)^2 + (1-5)^2} = 5$. This method is usually less expeditious than that of Art. 38.

EXAMPLES. Determine the length of the perpendicular from the point to the line in the following cases, ascertaining in each case whether the point and the origin are on the same or opposite sides of the line.

1. $3x + 4y - 2 = 0$, $(2, 7)$.

    Ans. $\frac{32}{5}$; on the opposite side from the origin.

2. $3x - 4y + 5 = 0$, $(8, 1)$.

    Ans. 5; on the side of the origin.

3. $4x - 3y - 6 = 0$, $(1, -1)$.

    Ans. $\frac{1}{5}$; on the opposite side from the origin.

4. $3x + 4y + 2 = 0$, $(2, 4)$.

    Ans. $\frac{24}{5}$; on the side of the origin.

5. $y - 2x + 1 = 0$, $(-1, -3)$.   *Ans.* 0.

6. Find the lengths of the altitudes of the triangle whose sides are $4x - 3y + 8 = 0$, $12x - 5y + 8 = 0$, $2x - y = 0$.

Ans. The vertices are $(1, 4)$, $(-4, -8)$, $(4, 8)$, and the altitudes $\frac{2}{\sqrt{5}}$, $\frac{16}{5}$, $\frac{16}{13}$.

7. Find the length of the altitudes of the triangle whose vertices are $(1, 2)$, $(-2, 0)$, $(6, -1)$.

Ans. $\frac{19}{\sqrt{13}}$, $\frac{19}{\sqrt{65}}$, $\frac{19}{\sqrt{34}}$.

58   ANALYTIC GEOMETRY.

8. Find the area of the triangle whose vertices are (2, 3), (−1, 4), (6, 5).

The line through (−1, 4) and (6, 5) is $x - 7y + 29 = 0$; its normal form is $\dfrac{-x}{\sqrt{50}} + \dfrac{7y}{\sqrt{50}} - \dfrac{29}{\sqrt{50}} = 0$. The distance of (2, 3) from this side is $\dfrac{-10}{\sqrt{50}}$. The length of the line joining (−1, 4) with (6, 5) is $\sqrt{50}$. Hence the area $= \frac{1}{2} \left(\dfrac{10}{\sqrt{50}} \times \sqrt{50} \right) = 5$.

9. Find the area of the triangle whose sides are $2x + y - 7 = 0$, $y - x - 1 = 0$, $x + 2y - 5 = 0$.

Ans. The vertices are (3, 1), (2, 3), (1, 2), and area $\frac{3}{2}$.

10. Find the distance between the parallels $y = 2x - 6$, $y = 2x + 8$.

The line $y = 2x - 6$ crosses Y at $(0, -6)$; the distance of this point from $y = 2x + 8$ is $-\dfrac{14}{\sqrt{5}}$.

11. Find the distance between the parallels
$$y = 3x, \quad y = 3x - 10. \qquad Ans. \ \sqrt{10}.$$

40. *To find the equation of a line bisecting the angle between two given lines.*

Let
$$x \cos a + y \sin a - p = 0, \qquad (1)$$
$$x \cos a' + y \sin a' - p' = 0, \qquad (2)$$

be the two given lines. Then
$$(x \cos a' + y \sin a' - p') + k(x \cos a + y \sin a - p) = 0 \qquad (3)$$

is a straight line through their intersection. Now the quantities in the parentheses are the distances of any point (x, y) from the lines (2) and (1) (Art. 38). Thus, if MN, $M'N'$, be the lines given by (1) and (2), then (3) is the equation of *some* line VP through their intersection V, and the parentheses are the distances PD', PD, of any of its points from $M'N$ and MN. Now if $k = -1$,

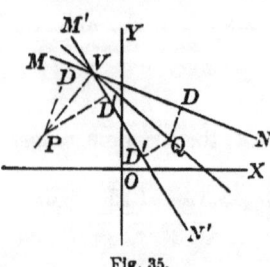

Fig. 35.

$PD' = PD$ from (3), and (3) will be the equation of the line bisecting the angle MVN'. When a point P is on the same side of a line as the origin, we have seen that the perpendicular PD is negative (Art. 38). For the angle MVN', P is on the same side of both lines that the origin is, and hence both perpendiculars must be negative, that is, have the same sign, or $k = -1$. For the angle $N'VN$, Q is on the same side of one line that the origin is, but on the opposite side from the origin in the case of the other line; one perpendicular must therefore be negative, and the other positive, that is, have opposite signs, or $k = 1$. Hence, *to bisect the angle between two given lines, put their equations under the normal form, and subtract or add them according as the origin does or does not lie within the angle to be bisected.*

EXAMPLES. 1. Find the bisector of the angle between $12x + 5y - 2 = 0$ and $3x - 4y + 7 = 0$, in which the origin lies. *Ans.* $21x + 77y - 101 = 0$.

2. Find the bisectors of the angles between $2x + y + 8 = 0$, $x + 2y - 3 = 0$. *Ans.* $3x + 3y + 5 = 0$; $x - y + 11 = 0$.

3. Find the bisectors of the angles between $2x + y + 8 = 0$, and $y = 0$. *Ans.* $2x + (1 \pm \sqrt{5})y + 8 = 0$.

4. Write the equations of the bisectors of the angles between the axes $y = 0$, $x = 0$. *Ans.* $y \pm x = 0$.

5. Of what line would Eq. (3), Art. 40, be the equation if $k = 2$? if $k = n$?

SECTION V. — THE POLAR SYSTEM.

41. *Derivation of polar from rectangular equations.* When the pole is taken at the origin and the polar axis is coincident with the axis of X, any rectangular equation of a straight line may be transformed into the corresponding polar equation (that is, the polar equation expressed in terms of the same constants) by means of the relations $x = r \cos \theta$, $y = r \sin \theta$ (Art. 23). The simplest and most useful of the polar equations is the normal form.

42. *Normal polar equation of the straight line.* The normal rectangular form being $x \cos a + y \sin a = p$, substituting $x = r \cos \theta$ and $y = r \sin \theta$, we have $r(\cos \theta \cos a + \sin \theta \sin a) = p$, whence

$$r = \frac{p}{\cos(\theta - a)}. \qquad (1)$$

DISCUSSION. If $\theta = 0°$, $r = \dfrac{p}{\cos(-a)} = \dfrac{p}{\cos a} = OQ$, locating the point Q where MN crosses the polar axes. If $\theta = a$, $r = \dfrac{p}{\cos 0°} = p$, giving the point D. If $\theta > a$ and increasing, $\theta - a$ is increasing, $\cos(\theta - a)$ decreases, and hence r increases till $\theta = a + 90°$, when $\cos(\theta - a) = \cos 90° = 0$ and $r = \infty$, as it should be, since r is then parallel to MN and must be produced infinitely to meet the line. When $\theta > a + 90°$, $\theta - a > 90°$, and r is negative, showing that it must be produced backwards, or away from the end of the measuring arc, to meet MN, and remains negative

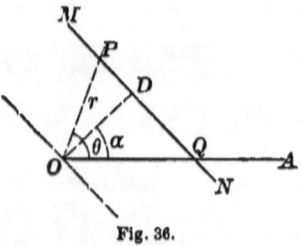

Fig. 36.

THE POLAR SYSTEM.

till $\theta = a + 270°$, or $\theta - a = 270°$, when $r = \infty$ again, and is parallel to MN. For $\theta = 360°$, $r = \dfrac{p}{\cos(-a)} = \dfrac{p}{\cos a} = OQ$. The entire line is traced for values of θ between 0° and 180°, for which latter value of θ, $r = \dfrac{p}{\cos(180° - a)} = -\dfrac{p}{\cos a} = OQ$.

If MN is perpendicular to the polar axis and lies on the right of the pole, $a = 0°$, and the equation becomes $r = \dfrac{p}{\cos \theta}$; if on the left of the pole, $a = 180°$, and the equation becomes

$$r = \dfrac{p}{\cos(\theta - 180°)} = \dfrac{p}{\cos -(180° - \theta)} = \dfrac{-p}{\cos \theta}.$$

Hence $r = \pm \dfrac{p}{\cos \theta}$ is the equation of all perpendiculars to the polar axis, the negative sign applying to those which lie on the left of the pole.

If the line is parallel to the polar axis and above it, $a = 90°$, and the equation becomes

$$r = \dfrac{p}{\cos(\theta - 90°)} = \dfrac{p}{\cos -(90° - \theta)} = \dfrac{p}{\sin \theta};$$

if below the polar axis, $a = 270°$, and

$$r = \dfrac{p}{\cos(\theta - 270°)} = \dfrac{p}{\cos -(270° - \theta)} = -\dfrac{p}{\sin \theta}.$$

Hence $r = \pm \dfrac{p}{\sin \theta}$ is the equation of all parallels to the polar axis, the negative sign applying to those which lie below the pole.

If the line passes through the pole, $p = 0$, and $r = 0$, except when $\theta = 90° + a$, in which case $r = \dfrac{0}{0}$; that is, r is zero for all values of θ except when the radius vector coincides with the line, when r may evidently have any value.

EXAMPLES. 1. Write the polar equation of a line whose distance from the pole is 5, the perpendicular being inclined 45° to the polar axis. Find the intercept on the axis, and the values of θ for which r is infinite.

$$\text{Ans. } r = \frac{5}{\cos(\theta - 45°)}; \quad 5\sqrt{2}; \quad 135°; \quad 315°.$$

2. Write the polar equations of lines for which $p = 2$, $\alpha = 60°$; $p = 10$, $\alpha = 120°$; and find their intercepts.

3. Construct $r = \dfrac{2}{\cos(\theta - 30°)}$; $r = \dfrac{8}{\cos(\theta - 60°)}$.

4. Construct $r = \pm \dfrac{4}{\cos\theta}$; $r = \pm \dfrac{5}{\sin\theta}$.

5. Write the polar equations of the sides of a square whose centre is at the pole and side 10, one side being parallel to the axis.

6. Find the rectangular equations of $r = \dfrac{3}{\cos\theta}$; $r = \dfrac{-5}{\sin\theta}$.

7. Find the rectangular equation of $r = \dfrac{9}{\cos(\theta - 45°)}$.

$$\text{Ans. } x + y - 9\sqrt{2} = 0.$$

8. Find the polar equation of $3x - 4y + 1 = 0$.

If the normal form is required, $\dfrac{3x - 4y + 1}{-5} = 0$, whence $p = \tfrac{1}{5}$, and $\alpha = \cos^{-1}\tfrac{3}{5}$, which may be found from the tables. Then substitute p and α in $r = \dfrac{p}{\cos(\theta - \alpha)}$. If the normal form is not specified, substituting directly the values of $x = r\cos\theta$ and $y = r\sin\theta$, we have $r = \dfrac{1}{4\sin\theta - 3\cos\theta}$.

9. Find the polar equation of $y = 3x + 2$.

$$\text{Ans. } r = \frac{2}{\sin\theta - 3\cos\theta}.$$

SECTION VI.—APPLICATIONS.

43. Recapitulation. The foregoing formulæ and equations relating to points and straight lines constitute the elementary tools, as it were, of analytic research on the properties of rectilinear figures. The student must remember that it is not the object of Analytic Geometry to produce these equations and formulæ, but *to investigate the properties of loci by means of them*. While, therefore, familiarity with these expressions is indispensable, a mastery of analytic geometry implies a knowledge of their *use* in the discovery of geometrical truths; that is, the mastery of a *method of research*. The more important of these expressions are here collected as a review exercise. The student should memorize them, and be able to explain the meaning of all the quantities involved. Thus, $x = \dfrac{x' + x''}{2}$, $y = \dfrac{y' + y''}{2}$, are the equations of a point midway between two given points, in which x, y, are the coordinates of the required middle point, and x', y', x'', y'', the coordinates of the given points.

$x = \dfrac{x' + x''}{2}$, $y = \dfrac{y' + y''}{2}$.   Equation (3), Art. 6.

$d = \sqrt{(x'' - x')^2 + (y'' - y')^2}$.   " (1), " 7.

$d = \sqrt{r'^2 + r''^2 - 2 r' r'' \cos(\theta'' - \theta')}$.   " (1), " 13.

$x = x_0 + x_1, \; y = y_0 + y_1$.   " (2), " 22.

$x = r \cos \theta, \; y = r \sin \theta$.   " (4), " 23.

$Ax + By + C = 0$.   " (1), " 25.

$\dfrac{x}{a} + \dfrac{y}{b} = 1$.   " (1), " 26.

$y = mx + b$.   " (2), " 26.

$x \cos a + y \sin a = p.$ Equation (3), Art. 26.

$y - y' = m(x - x').$ " (2), " 31.

$y - y' = \dfrac{y' - y''}{x' - x''}(x - x').$ " (4), " 32.

$c = \dfrac{m' - m}{1 + mm'}.$ " (1), " 33.

$y - y' = \dfrac{m + c}{1 - mc}(x - x').$ " (2), " 34.

$m = m', \quad m = -\dfrac{1}{m'}.$ " 35.

$y - y' = m(x - x').$ " (1), " 35.

$y - y' = -\dfrac{1}{m}(x - x').$ " (2), " 35.

$D = \dfrac{Ax' + By' + C}{\sqrt{A^2 + B^2}}.$ " (1), " 38.

$x \cos a' + y \sin a' - p' \pm (x \cos a + y \sin a - p) = 0.$ " 40.

$r = \dfrac{p}{\cos(\theta - a)}.$ " (1), " 42.

PROPERTIES OF RECTILINEAR FIGURES.

44. 1. *The diagonals of a square are perpendicular to each other.*

Take two adjacent sides for the axes. Then, if $a =$ side, the vertices are $(0, 0)$, $(a, 0)$, (a, a), $(0, a)$, and the equations of the diagonals are $y = x$, $y = -x + a$, in which $m = -\dfrac{1}{m'}$ (Art. 35).

2. *The line joining the middle points of two sides of a triangle is parallel to the third side.*

Take the third side for the axis of X, and the origin at its left-hand extremity. Then $(0, 0)$, $(a, 0)$, (b, c) are the vertices, $\left(\dfrac{b}{2}, \dfrac{c}{2}\right)$, $\left(\dfrac{a+b}{2}, \dfrac{c}{2}\right)$, the middle points, and $y = \dfrac{c}{2}$ is the line joining them.

APPLICATIONS.

3. *The diagonals of a parallelogram bisect each other.*

With the axes as in the figure, let the side $OB = a$, the altitude $mD = b$, and $Om = c$. Then the coordinates of C are $(a + c,\ b)$, and the middle point of OC is $\left(\dfrac{a+c}{2},\ \dfrac{b}{2}\right)$. The coordinates of B are $(a,\ 0)$, of D, $(c,\ b)$, and of the middle point of DB, $\left(\dfrac{a+c}{2},\ \dfrac{b}{2}\right)$.

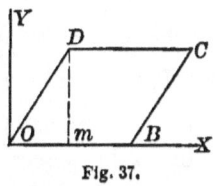

Fig. 37.

4. *The straight lines joining the middle points of the opposite sides of any quadrilateral bisect each other.*

Let $(0,\ 0)$, $(a,\ 0)$, $(b,\ c)$, $(d,\ e)$ be the vertices $O,\ B,\ C,\ D$, in order. Then the middle point of each line is

$$\left(\frac{a+b+d}{4},\ \frac{c+e}{4}\right).$$

5. *Prove that the middle point of the line joining the middle points of the diagonals of any quadrilateral is the point of intersection of the lines of Ex. 4.*

6. *The lines joining the middle points of the adjacent sides of a parallelogram form a parallelogram.*

With the notation of Ex. 3, the slope of the lines joining the middle points of DC and BC, DO and OB, is $\dfrac{b}{c-a}$; hence these lines are parallel.

7. *The middle point of the hypothenuse of any right-angled triangle is equally distant from the vertices.*

Take the axes coincident with the sides.

8. *Prove that if A, B, C, be squares on the sides of a right-angled triangle ORQ, and OT is perpendicular to RQ, then RS, QP, and OT meet in a point.* With the axes as in the figure, let $c = OQ$, $d = OR$, the sides. Then the coordinates of S and R are $(c,\ -c)$, $(-d,\ 0)$, and the equation of SR is

$y = -\dfrac{c}{c+d}x - \dfrac{cd}{c+d}$. Similarly the equation of QP is $y = -\dfrac{c+d}{d}x - c$. The equation of RQ is $\dfrac{x}{-d} + \dfrac{y}{-c} = 1$, and of OT, perpendicular to it, $y = \dfrac{d}{c}x$. Substituting this value of y in the equations of RS and QP, the values of x are found to be the same; hence OT intersects them both at the same point.

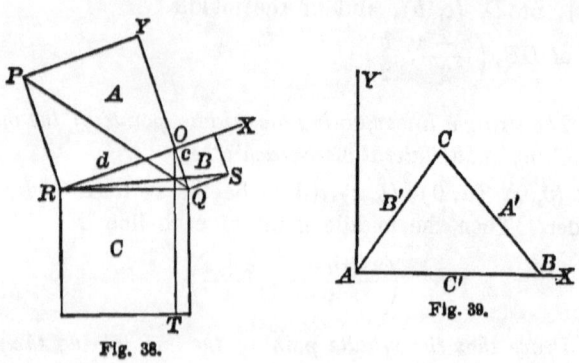

Fig. 38.   Fig. 39.

9. *The altitudes of a triangle meet in a point.*

Take the axes as in the figure, and let $AB = c$, C being given as (x', y'). Then the altitude through C is

$$x = x'. \qquad (1)$$

The equation of BC is $y - y' = \dfrac{y'}{x'-c}(x-x')$, and that of the altitude through A is

$$y = \dfrac{c-x'}{y'}x. \qquad (2)$$

The equation of AC is $y = \dfrac{y'}{x'}x$, and of the altitude through B is

$$y = -\dfrac{x'}{y'}(x-c). \qquad (3)$$

Combining (2) and (3) to find their intersection, we obtain $x = x'$, which satisfies (1). Hence (1), (2), and (3) meet in a point.

APPLICATIONS.

10. *The perpendiculars erected at the middle points of the sides of a triangle meet in a point.*

The equation of AC (Fig. 39) is $y = \dfrac{y'}{x'}x$, and that of the perpendicular to AC through B' is

$$y - \frac{y'}{2} = -\frac{x'}{y'}\left(x - \frac{x'}{2}\right). \tag{1}$$

The equation of BC is $y - y' = \dfrac{y'}{x' - c}(x - x')$, and that of the perpendicular to BC through A' is

$$y - \frac{y'}{2} = \frac{c - x'}{y'}\left(x - \frac{c + x'}{2}\right). \tag{2}$$

The perpendicular to AB at C' is

$$x = \frac{c}{2}. \tag{3}$$

Combining (1) and (2), eliminating y, we have $x = \dfrac{c}{2}$. Hence (1) and (2) intersect on (3).

11. *The medials of a triangle meet in a point.*

The middle points A', B', C' (Fig. 39), are

$$\left(\frac{c + x'}{2}, \frac{y'}{2}\right), \quad \left(\frac{x'}{2}, \frac{y'}{2}\right), \quad \left(\frac{c}{2}, 0\right),$$

and the equations of the medials are

$$y = \frac{y'}{c + x'}x, \quad AA', \tag{1}$$

$$y = \frac{y'}{x' - 2c}(x - c), \quad BB', \tag{2}$$

$$y = \frac{y'(2x - c)}{2x' - c}, \quad CC'. \tag{3}$$

Combining (2) and (3), we find they intersect in $\left(\dfrac{x' + c}{3}, \dfrac{y'}{3}\right)$, and these values satisfy (1); hence (1), (2), and (3) meet in a point.

68   ANALYTIC GEOMETRY.

12. *To find the general analytic condition that three straight lines may meet in a point.*

Let (1) $y = m'x + b'$, (2) $y = m''x + b''$, (3) $y = m'''x + b'''$ be the three lines. The intersection of (1) and (2) is

$$x = \frac{b'' - b'}{m' - m''}, \quad y = \frac{m'b'' - b'm''}{m' - m''}.$$

But these must satisfy (3); hence

$$m'b'' - m''b' + m'''b' - m'b''' + m''b''' - m'''b'' = 0.$$

13. *Show that*

$$11y - 20x - 8 = 0, \quad y - 4x = 0, \quad 13y - 28x - 8 = 0,$$

meet in a point.

14. *To find an expression for the area of a triangle in terms of the coordinates of its vertices.*

Let (x', y'), (x'', y''), (x''', y'''), be the vertices. The equation of a line through the first two is $y - y' = \frac{y' - y''}{x' - x''}(x - x')$, or $(y'' - y')x + (x' - x'')y + y'x'' - y''x' = 0$. Hence the perpendicular distance from this side to (x''', y''') is

$$\frac{(y'' - y')x''' + (x' - x'')y''' + y'x'' - y''x'}{\sqrt{(y' - y'')^2 + (x' - x'')^2}}.$$

But the denominator of this expression is the distance between (x', y') and (x'', y''). Hence the area

$$= \tfrac{1}{2} \text{base} \times \text{altitude} = \tfrac{1}{2}[x'(y''' - y'') + x''(y' - y''') + x'''(y'' - y')].$$

15. *Find the area of the triangle whose vertices are* (2, 3), (−1, 4), (6, 5).

16. *Find the equation of a straight line passing through a given point and dividing the line joining two given points in a given ratio.*

Substitute in $y - y' = \frac{y' - y''}{x' - x''}(x - x')$ for x'', y'', the values of

APPLICATIONS. 69

x and y in Equation (2), Art. 6, and for x', y', the coordinates h, k, of the given point, and we have

$$y - k = \frac{m(y''-k) + n(y'-k)}{m(x''-h) + n(x'-h)}(x-h).$$

17. *The bisectors of the interior angles of a triangle meet in a point.*

Let the equations of the sides of the triangle be

$$x \cos a' + y \sin a' - p' = 0, \qquad (1)$$

$$x \cos a'' + y \sin a'' - p'' = 0, \qquad (2)$$

$$x \cos a''' + y \sin a''' - p''' = 0, \qquad (3)$$

and let the origin be *within* the triangle. Then the origin lies within each of the three angles to be bisected, and the equations of the bisectors (Art. 40) are

$$x \cos a' + y \sin a' - p' - (x \cos a'' + y \sin a'' - p'') = 0, \quad (4)$$

$$x \cos a'' + y \sin a'' - p'' - (x \cos a''' + y \sin a''' - p''') = 0, \quad (5)$$

$$x \cos a''' + y \sin a''' - p''' - (x \cos a' + y \sin a' - p') = 0. \quad (6)$$

But values of x and y which satisfy any two of these equations also satisfy the third; hence these three lines meet in a point.

18. *The lines which pass through the vertices of a triangle and bisect the angles supplemental to those of the triangle meet in a point.*

For brevity, represent by $a = 0$, $\beta = 0$, $\gamma = 0$, Equations (1), (2), and (3) of Ex. 17. Then $a + \beta = 0$, $\beta + \gamma = 0$, $\gamma + a = 0$ are the bisectors required.

19. *The bisectors of any two exterior angles of a triangle and of the third interior angle meet in a point.*

The bisector of the exterior angle of (1) and (2), Ex. 17, is $a + \beta = 0$, and of (2) and (3) is $\beta + \gamma = 0$, and the bisector of the interior angle of (1) and (3) is $a - \gamma = 0$. Subtracting the second of these equations from the first, we have the third.

20. *The bisectors of the angles between the bisectors of perpendiculars are the perpendiculars themselves.*

Let $y = mx + b$, $y = -\dfrac{1}{m}x + b'$, be the perpendiculars. Their normal forms are

$$\frac{y - mx - b}{\sqrt{1 + m^2}} = 0, \qquad \frac{my + x - mb'}{\sqrt{1 + m^2}} = 0,$$

and their bisectors are $y - mx - b \pm (my + x - mb') = 0$. The normal forms of these latter are

$$\frac{(1 + m)y + (1 - m)x - (mb' + b)}{\sqrt{(1 + m)^2 + (1 - m)^2}} = 0,$$

$$\frac{(1 - m)y - (1 + m)x + (mb' - b)}{\sqrt{(1 + m)^2 + (1 - m)^2}} = 0,$$

and their bisectors are

$$(1 + m)y + (1 - m)x - (mb' + b)$$
$$\pm [(1 - m)y - (1 + m)x + (mb' - b)] = 0,$$

or $y - mx - b = 0$, and $my + x - mb' = 0$, which are the given perpendiculars.

21. *To find the condition that the three points (x', y'), (x'', y''), (x''', y'''), shall be collinear, i.e., lie on the same straight line.*

22. *Prove that the line which divides two sides of a triangle proportionally is parallel to the third side.*

CHAPTER III.

EQUATION OF THE SECOND DEGREE. THE CONIC SECTIONS.

SECTION VII.—COMMON EQUATIONS OF THE CONIC SECTIONS.

45. The Conic Sections. It has been shown in the previous chapter that every complete equation of the *first* degree between x and y, $Ax + By + C = 0$, and the various forms which such an equation may assume owing to a change in the values or signs of the arbitrary constants A, B, C, is the equation of a straight line. In the present chapter it will be shown that every equation of the *second* degree between x and y, $Ay^2 + Bxy + Cx^2 + Dy + Ex + F = 0$, and the various forms it may assume when different values and signs are given to the arbitrary constants A, B, C, D, E, F, represents some one of a family of loci called the **Conic Sections**. These loci, which for brevity may be designated Conics, are so named because every section of the surface of a right cone with a circular base by a plane is one of this family.

They may all be traced by a point so moving that *the ratio of its distances from a fixed point and a fixed straight line remains constant*, the particular locus traced depending upon the value of this constant. Since all the loci of this family may thus be generated by a point moving under a single law, it will evidently be possible to express this law in a single equation, and to derive the particular cases from this general equation by assigning the corresponding value to the ratio. The proof of the foregoing statements and the discussion of the general equation is, however, greatly facilitated by a knowledge of the forms and elementary properties of these loci; we shall therefore first determine their equations separately from some of their properties with a view to the discovery of their forms, reserving the discussion of the general equation until the student has thus become familiar with the various loci which it represents.

THE CIRCLE.

46. Defs. The path of a point so moving that its distance from a fixed point remains constant is a **circle**. The constant distance is the **radius**, the fixed point the **centre**.

47. General equation of the circle.

Let (m, n) be the centre C, R the radius, and P any point of the circle. From the right-angled triangle PCM, $CP^2 = CM^2 + MP^2$,

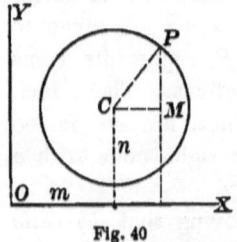

Fig. 40

or $\quad (y-n)^2 + (x-m)^2 = R^2, \qquad (1)$

which is the required equation. Hence, *to write the equation of any circle whose position and radius are known, substitute the given values of m, n, and R, in the above equation.* Thus, the equation of the circle whose radius is 6 and centre is $(6, -2)$, is $(y+2)^2 + (x-6)^2 = 36$, or $y^2 + x^2 + 4y - 12x + 4 = 0$.

By assigning different values to m and n, we may derive the equation of a circle in any position from the general equation (1). Two of these derived equations are of frequent use and should be memorized. First: when the centre is at the origin, in which case $m = 0$, $n = 0$, and (1) becomes

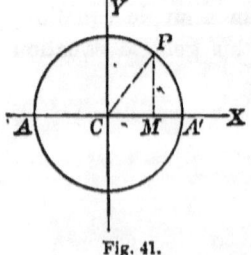

Fig. 41.

$$y^2 + x^2 = R^2, \qquad (2)$$

called the **central** equation of the circle. Second: when the origin is at the left-hand extremity of any diameter assumed as the axis of X, in which case $m = R$, $n = 0$, and (1) becomes

$$y^2 = 2Rx - x^2. \qquad (3)$$

Thus, the central equation of the circle whose radius is 6 is

$$y^2 + x^2 = 36,$$

and when referred as in Fig. 42,

$$y^2 = 12x - x^2.$$

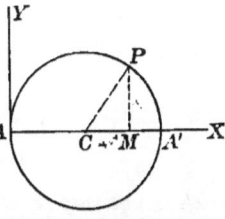

Fig. 42.

The above forms may be obtained directly from the corresponding figures. The student will observe that, by transposition, either (2) or (3) shows that $PM^2 = AM \cdot MA'$, a well-known property of the circle from which these equations might have been established.

48. *The equation of every circle is some form of the equation*

$$y^2 + x^2 + Dy + Ex + F = 0. \qquad (1)$$

Expanding the general equation of the circle

$$(y - n)^2 + (x - m)^2 = R^2, \qquad (2)$$

we have $\quad y^2 + x^2 - 2ny - 2mx + m^2 + n^2 - R^2 = 0, \qquad (3)$

which is of the same form as (1). The first two terms of (3) are independent of m, n, and R, so that no change in the position or magnitude of the circle can affect these terms. *Every equation of a circle, therefore, will contain the squares of x and y with equal coefficients and like signs.* The remaining terms will vary with the radius and position of the circle. Thus $E = 0$, when $m = 0$, and $y^2 + x^2 + Dy + F = 0$, is the equation of all circles whose centres are on Y; $D = 0$, when $n = 0$, and $y^2 + x^2 + Ex + F = 0$ applies to all circles whose centres are on X; if both m and n are zero, than $E = 0$, $D = 0$, and we have the central form $y^2 + x^2 = R^2$, the centre being at the origin; if the origin is on the curve, then the equation can have no absolute term, or $F=0$, and (1) becomes $y^2 + x^2 + Dy + Ex = 0$; if D, E, and F are all zero, we have $y^2 + x^2 = 0$, which is true only for $x=0$, $y=0$, the origin, the circle becoming a point; this may be regarded as the limiting case of the central form as the radius diminishes indefinitely.

49. *Conversely, every equation of the form*
$$y^2 + x^2 + Dy + Ex + F = 0, \tag{1}$$
which is not impossible, is the equation of a circle.

Adding to both members of (1) the squares of half the coefficients of x and y, we obtain

$$y^2 + Dy + \frac{D^2}{4} + x^2 + Ex + \frac{E^2}{4} = \frac{D^2}{4} + \frac{E^2}{4} - F,$$

or $\quad\left(y + \frac{D}{2}\right)^2 + \left(x + \frac{E}{2}\right)^2 = \tfrac{1}{4}(D^2 + E^2 - 4F), \tag{2}$

which is of the same form as

$$(y - n)^2 + (x - m)^2 = R^2, \tag{3}$$

and in which, therefore,

$$-\frac{D}{2} = n, \quad -\frac{E}{2} = m, \quad \tfrac{1}{4}(D^2 + E^2 - 4F) = R^2. \tag{4}$$

If $D^2 + E^2 > 4F$, then R^2 is positive, R is real, and the equation represents a circle whose centre is $\left(-\frac{E}{2}, -\frac{D}{2}\right)$, and whose radius is $\tfrac{1}{2}\sqrt{D^2 + E^2 - 4F}$. If $D^2 + E^2 = 4F$, then R^2 is zero, and the equation becomes $\left(y + \frac{D}{2}\right)^2 + \left(x + \frac{E}{2}\right)^2 = 0$, which is satisfied only for $x = -\frac{E}{2}$, $y = -\frac{D}{2}$, or the circle becomes a point, namely, the centre, which may be regarded as a circle whose radius is zero. If $D^2 + E^2 < 4F$, then R^2 is negative, R is imaginary, and the equation is impossible since the sum of two squares cannot be negative. We have thus three cases, in which R is real, zero, or imaginary, and for brevity and

COMMON EQUATIONS OF CONIC SECTIONS. 75

uniformity of expression we may say that *every equation of the form $y^2 + x^2 + Dy + Ex + F = 0$ is the equation of a circle, real or imaginary.*

The equation $ay^2 + ax^2 + dy + ex + f = 0$ may be reduced to the form of (1), and is therefore the most general form which the equation of a circle can assume.

50. *To determine the centre and radius of a circle whose equation is given.*

When the equation is given in the form $(y-n)^2 + (x-m)^2 = R^2$, the centre (m, n) and radius R may, of course, be determined by inspection. If given in the form $y^2 + x^2 + Dy + Ex + F = 0$, we may put it under the above form by *adding to both members the squares of half the coefficients of x and y,* as in Art. 49. Otherwise, by equations (4), Art. 49, *the coordinates of the centre are half the coefficients of x and y with their signs changed, and the radius* $R = \frac{1}{2}\sqrt{D^2 + E^2 - 4F}$. Thus, given $y^2 + x^2 - 4y + 2x + 1 = 0$, $m = -1$, $n = 2$, $R = \frac{1}{2}\sqrt{16 + 4 - 4} = 2$.

51. *The equations of concentric circles differ only in their absolute terms.*

Since the values of $m\left(=-\frac{E}{2}\right)$ and $n\left(=-\frac{D}{2}\right)$ are independent of F, and $R = \frac{1}{2}\sqrt{D^2 + E^2 - 4F}$, if in the equation of any circle F changes, D and E remaining the same, the circle retains its position but changes its size. Hence *circles are concentric whose equations differ only in their absolute terms.*

EXAMPLES. 1. Write the equation of the circle whose radius is 7 and centre at $(0, 8)$.

Ans. $(y-8)^2 + (x-0)^2 = 49$, or $y^2 + x^2 - 16y + 15 = 0$.

2. Write the equations of the following circles:

Centre at $(-1, -4)$, radius 2;
Centre at $(0, 0)$, radius 9;
Centre at $(5, 0)$, radius 5;
Centre at $(-5, 5)$, radius 5.

3. Write the equation of a circle whose centre is (6, 8), passing through the origin.

4. Which of the following equations are those of circles?

$x^2 + 2y^2 + 8x - 4y + 7 = 0$; $x^2 + y^2 + xy + x + y - 1 = 0$;
$x^2 - y^2 + x - 2y + 4 = 0$; $y^2 + x^2 - 4y - 8x + 1 = 0$;
$x^2 - 3y^2 + x - 8y = 0$; $2x^2 + 2y^2 + 4x - 3y + 7 = 0$;
$x^2 + 2y - 4x - 1 = 0$; $\frac{3}{5}y^2 + \frac{3}{5}x^2 - 2x = 0$.

5. Write the equations of a circle whose radius is 6 when (1) both axes are tangent to the circle; (2) when $X'X$ is a tangent and $Y'Y$ passes through the centre (two cases).

6. Determine the position and radius of the following circles:

$y^2 + x^2 - 8y + 4x - 5 = 0$.   Ans. $(-2, 4)$, 5.
$y^2 + x^2 + 10y - 4x - 7 = 0$.   Ans. $(2, -5)$, 6.
$y^2 + x^2 + 10y + 4x - 20 = 0$.   Ans. $(-2, -5)$, 7.
$y^2 + x^2 - 2y + 6x = 0$.   Ans. $(-3, 1)$, $\sqrt{10}$.
$y^2 + x^2 + 3y - 7x - \frac{3}{2} = 0$.   Ans. $(\frac{7}{2}, -\frac{3}{2})$, 4.
$y^2 + x^2 + 4y - 2x + 5 = 0$.   Ans. $(1, -2)$, 0.
$36y^2 + 36x^2 - 24y - 36x - 131 = 0$. Ans. $(\frac{1}{2}, \frac{1}{3})$, 2.
$y^2 + x^2 + y + x - 1 = 0$.   Ans. $(-\frac{1}{2}, -\frac{1}{2})$, $\frac{1}{2}\sqrt{6}$.
$y^2 + x^2 + y + x + 1 = 0$.   Ans. $(-\frac{1}{2}, -\frac{1}{2})$, $\frac{1}{2}\sqrt{-2}$.
$y^2 + x^2 - y - x + 4 = 0$.   Ans. $(\frac{1}{2}, \frac{1}{2})$, $\frac{1}{2}\sqrt{-14}$.

7. Write the equation of the circle whose centre is at the origin, and which touches the line $3x - 4y + 25 = 0$.

Putting the equation of the line under the normal form,

$$-\frac{3x}{5} + \frac{4y}{5} = 5 = p,$$

which must equal R. Hence $y^2 + x^2 = 25$.

8. Write the equation of the circle whose centre is (0, 0), and which touches the line $3x + y - 6 = 0$.

9. Write the equation of the circle whose centre is (2, 3), and which touches $3x + 4y + 12 = 0$.

COMMON EQUATIONS OF CONIC SECTIONS.

10. Prove that the sum of the equations of any number of circles is the equation of a circle.

11. Prove that if the equation of a straight line be added to the equation of a circle, the sum is the equation of a circle.

52. Polar equation of the circle.

The general equation of the circle being $(y-n)^2+(x-m)^2=R^2$, let the pole be taken at the origin and the polar axis coincident with X. Then, if r', θ' (Fig. 43), are the polar coordinates of the centre C, and r, θ, those of any point P, from the formulæ for transformation, Eq. (4), Art. 23, we have $x=r\cos\theta$, $y=r\sin\theta$, $m=r'\cos\theta'$, $n=r'\sin\theta'$, which being substituted in the above equation, there results, after reduction,

$$r^2 - 2rr'\cos(\theta-\theta') = R^2 - r'^2. \qquad (1)$$

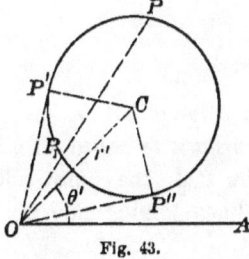
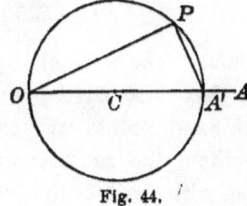

Fig. 43.   Fig. 44.

From this equation we may derive that of the circle in *any* given position by assigning the proper values to r' and θ'. Thus, if the centre is at the pole, $r'=0$, and (1) becomes

$$r = R, \qquad (2)$$

which is true for all values of θ. (See Ex. 1, Art. 19.) If the pole is on the curve, and the polar axis a diameter, $\theta'=0$, $r'=R$, and (1) becomes

$$r = 2R\cos\theta, \qquad (3)$$

which, being true for all positions of P (Fig. 44), shows that OPA', or *the angle inscribed in a semi-circle, is a right angle.* (See Ex. 2, Art. 19.)

DISCUSSION OF EQUATION (1). Solving the equation for r, we have

$$r = r' \cos(\theta - \theta') \pm \sqrt{R^2 - r'^2 \sin^2(\theta - \theta')},$$

which gives two values of r for every value of θ, locating two points P and P_1, so long as $R^2 > r'^2 \sin^2(\theta - \theta')$, or $R > r' \sin(\theta - \theta')$. If $R < r' \sin(\theta - \theta')$, r is imaginary. If $R = r' \sin(\theta - \theta')$, r has but one value; in this case the two points P and P_1 coincide, the secant OP becoming the tangent OP', or OP'', and $r = r' \cos(\theta - \theta') = OP' = OP''$. Since this relation is true only when the triangles OCP' and OCP'' are right triangles, we see that *the radius is perpendicular to the tangent at the point of contact;* this also appears from the condition $R = r' \sin(\theta - \theta')$, which must be fulfilled when r has but a single value, this condition being $CP' = OC \sin COP'$, or $CP'' = OC \sin COP''$.

EXAMPLES. 1. Write the polar equation of the circle whose radius is 10, the pole being on the circle and the polar axis a diameter. Discuss the equation, showing that the entire circle is traced for values of θ from 0° to 180°.

2. Construct the circles whose equations are $r = 8 \cos\theta$, $r = -8 \cos\theta$.

3. Derive the polar form $r = 2R \cos\theta$ from the corresponding rectangular form $y^2 = 2Rx - x^2$.

THE ELLIPSE.

53. Defs. *The path of a point so moving that the sum of its distances from two fixed points is constant is called an* **ellipse.** The two fixed points are called the **foci**, the point midway between them the **centre**, and the lines joining any point of the ellipse with the foci the **focal radii**.

54. Central equation of the ellipse.

Let F, F' be the foci, O the centre and origin, the axis of X being coincident with FF', P any point of the ellipse, $FF' = 2c$, and $2a$ the constant sum. Then $FP + F'P = 2a$. But

$$FP = \sqrt{PM^2 + MP^2} = \sqrt{(x+c)^2 + y^2},$$
$$F'P = \sqrt{F'M^2 + MP^2} = \sqrt{(x-c)^2 + y^2}.$$

Hence $\sqrt{(x+c)^2 + y^2} + \sqrt{(x-c)^2 + y^2} = 2a$.

Transposing the second term to the second member, and squaring,

$$(x+c)^2 + y^2 = 4a^2 - 4a\sqrt{(x-c)^2 + y^2} + (x-c)^2 + y^2,$$

or $\qquad cx - a^2 = -a\sqrt{(x-c)^2 + y^2}.$

COMMON EQUATIONS OF CONIC SECTIONS.

Squaring again,
$$a^2y^2 + (a^2 - c^2)x^2 = a^2(a^2 - c^2). \tag{1}$$

DISCUSSION OF THE EQUATION. Since only the squares of the variables enter the equation, *the ellipse is symmetrical with respect to both axes.* Making $y = 0$, the X-intercepts are $\pm a$. Take $OA = OA' = a$, then $AA' = 2a =$ the constant sum, and as the sum of two sides of a triangle is greater than the third side, $PF + PF' = 2a > FF' = 2c$, or A and A' are *without the*

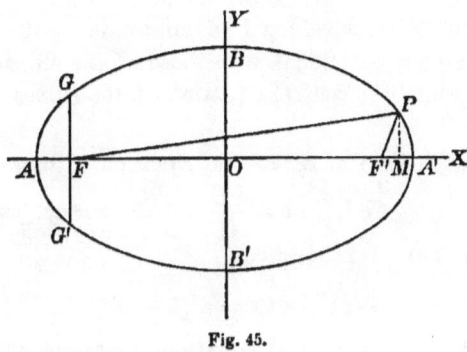

Fig. 45.

foci. Making $x = 0$, the Y-intercepts are $\pm \sqrt{a^2 - c^2}$, which are real, since $a > c$, and locate the points B, B'. Solving the equation for y,
$$y = \pm \frac{1}{a} \sqrt{(a^2 - c^2)(a^2 - x^2)},$$
which is imaginary if $x > a$ numerically, and therefore *the curve lies wholly within the limits A and A' along* X. Solving for x,
$$x = \pm a \sqrt{1 - \frac{y^2}{a^2 - c^2}},$$
which is imaginary if $\frac{y^2}{a^2 - c^2} > 1$, or $y > \sqrt{a^2 - c^2}$ numerically, or *the curve lies wholly within the limits B and B' along* Y. The form of the ellipse is best observed by the following mechanical construction: Take a string whose length is $AA' = 2a$, fix its extremities at F and F', place a pencil point against the

string, keeping the string stretched; as the pencil is moved it will trace the ellipse, for in all its positions $FP + PF' = 2a$.

55. Defs. AA' is called the **transverse axis** of the ellipse, BB' the **conjugate axis**, A and A' the **vertices**, FA and FA' (or $F'A$ and $F'A'$) the **focal distances**, the double ordinate through either focus, as GG', the **parameter**, and the distance from the focus to the centre divided by the semi-transverse axis $\left(\dfrac{FO}{AO}\right)$ the **eccentricity**. As referred to an origin at its centre and axes of reference coincident with those of the ellipse, Eq. (1), Art. 54, is called the **central equation** of the ellipse.

56. Common form *of the central equation.* Representing the eccentricity by e, we have $e = \dfrac{FO}{AO} = \dfrac{c}{a}$, $\therefore c = ae$, which substituted in Eq. (1), Art. 54, gives

$$y^2 + (1 - e^2)x^2 = a^2(1 - e^2), \qquad (1)$$

another form of the central equation, in terms of the eccentricity. Representing the conjugate axis BB' by $2b$, $2b = 2\sqrt{a^2 - c^2}$, $\therefore a^2 - c^2 = b^2$, which substituted in Eq. (1), Art. 54, gives

$$a^2y^2 + b^2x^2 = a^2b^2, \qquad (2)$$

the equation of the ellipse in terms of the semi-axes, and called the **common form** of the central equation.

Cor. 1. Since $e = \dfrac{c}{a}$ and $a > c$, *the eccentricity of the ellipse is always less than unity.*

Cor. 2. Since $c^2 = a^2 - b^2$, $e = \dfrac{c}{a} = \dfrac{\sqrt{a^2 - b^2}}{a}$, *the eccentricity in terms of the semi-axes.*

Cor. 3. Since $e = \dfrac{c}{a}$, $c = ae$, *the distance of either focus from the centre.*

NOTE. The student will observe that the form of the ellipse will vary with a, b, c, and e, and therefore that the constants in the equation of a locus may serve to determine its form as well as its magnitude and position (Art. 16).

57. Length of the focal radii.

P being any point of the ellipse (Fig. 45),

$$FP^2 = FM^2 + MP^2 = (ae + x)^2 + y^2 \text{ (Art. 56, Cor. 3)}$$
$$= (ae + x)^2 + (a^2 - x^2)(1 - e^2) \text{ (Art. 56, Eq. 1)}$$
$$= a^2 + 2aex + e^2x^2 = (a + ex)^2;$$

or $\quad FP = a + ex.$

But $\quad FP + F'P = 2a, \therefore F'P = 2a - (a + ex) = a - ex.$

Hence, *the focal radii to any point whose abscissa is x are $a \pm ex$.*

58. Polar equation of the ellipse.

Let the pole be taken at the left-hand focus, and the polar axis coincident with the transverse axis. We shall obtain the equation directly from the figure, this being easier than to transform the central equation. From the triangle FPF', P being any point of the curve,

$$F'P^2 = FP^2 + FF'^2 - 2FP \cdot FF' \cos F'FP.$$

But $FP = r$, $F'FP = \theta$, $FF' = 2ae$, and $F'P = 2a - FP = 2a - r$. Making these substitutions, we obtain

$$r = \frac{a(1 - e^2)}{1 - e \cos \theta}. \tag{1}$$

DISCUSSION OF THE EQUATION. When

$$\theta = 0°, \; r = a(1 + e) = FA';$$

when $\theta = 180°$, $r = a(1 - e) = FA$; hence *the focal distances are* $a(1 \pm e)$.

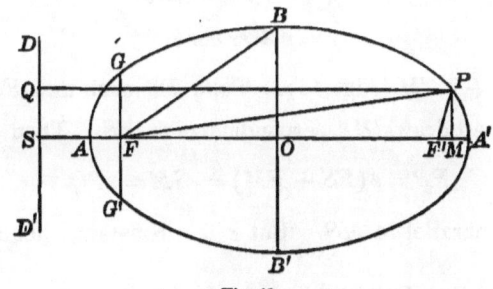

Fig. 46.

When $\theta = 90°$, $r = a(1-e^2) = a\left(1 - \dfrac{c^2}{a^2}\right) = a\dfrac{a^2-c^2}{a^2} = \dfrac{b^2}{a}$;

hence *the parameter* $GG' = 2a(1-e^2)$, or $\dfrac{2b^2}{a}$.

When $\theta = F'FB = \cos^{-1}\dfrac{FO}{FB} = \cos^{-1}\dfrac{ae}{r}$, $r = \dfrac{a(1-e^2)}{1 - \dfrac{ae^2}{r}}$,

$\therefore r = a = FB$. This is also evident from the right-angled triangle FOB, in which $FO = c$, $OB = b$, and therefore

$$FB = \sqrt{c^2 + b^2} = a,$$

since $a^2 - c^2 = b^2$. Hence *the distance from either focus to the extremity of the conjugate axis is equal to the semi-transverse axis.* Therefore, *to find the foci when the axes are given, with the extremity of the conjugate axis as a centre and the semi-transverse axis as a radius describe an arc; it will cut the transverse axis in the foci.*

Representing the parameter $GG' = 2a(1-e^2)$ by $2p$, the polar equation (1) may be written

$$r = \dfrac{p}{1 - e\cos\theta}. \qquad (2)$$

59. The ratio. *The ellipse can be traced by a point so moving that the ratio of its distances from a fixed point and a fixed straight line is constant.*

From the polar equation $r = \dfrac{p}{1 - e\cos\theta}$, we have

$$r = p + er\cos\theta,$$

or $FP = p + 2\,FM$ (Fig. 46). Take FS such that $FS = \dfrac{p}{e}$, or $p = eFS$, and draw DD' perpendicular to FS. Then

$$FP = e(FS + FM) = eSM = ePQ,$$

PQ being parallel to MS. But e is a constant; hence $\dfrac{FP}{PQ}$ is a constant.

The fixed line DD' is called the **Directrix**.

COMMON EQUATIONS OF CONIC SECTIONS. 83

Cor. 1. *The ratio is equal to the eccentricity and is always less than unity.*

Cor. 2. Since A is a point of the curve, $\dfrac{AF}{AS} = e$,

$$\therefore AS = \frac{AF}{e} = \frac{a(1-e)}{e} \text{ (Art. 58).}$$

For the same reason $\dfrac{A'F}{A'S} = e$, $\therefore A'S = \dfrac{A'F}{e} = \dfrac{a(1+e)}{e}$. Hence, the distances from the vertices to the directrix are $\dfrac{a(1 \mp e)}{e}$.

Cor. 3. $FS = FA + AS = a(1-e) + \dfrac{a(1-e)}{e} = \dfrac{a(1-e^2)}{e}$, the distance from the focus to the directrix.

Cor. 4. $OS = OF + FS = ae + \dfrac{a(1-e^2)}{e} = \dfrac{a}{e}$, the distance from the centre to the directrix.

60. Geometrical construction *of the ellipse when the ratio is given.*

Let $e = \dfrac{k}{s}$, in which $k < s$, be the given ratio. Take $SF = s$, draw GG' perpendicular to SF, and make $FG = FG' = k$. Draw SG and SG', and between these lines produced draw any parallel to GG', as $N'L'$. With F as a centre and $M'N'$ as a radius describe an arc cutting the parallel in P' and P''; these are points of the ellipse. To prove that P' is a point of the ellipse, draw DD' perpendicular to SF through S, and $P'Q'$ parallel to FS. Then, from similar triangles,

$$N'M' : M'S :: GF : FS;$$

but $\qquad N'M' = FP',\ M'S = P'Q'$;

hence $\qquad FP' : P'Q' :: GF : FS,$

or $\qquad \dfrac{FP'}{P'Q'} = \dfrac{GF}{FS} = e.$

In the same way any number of points may be constructed.

84   ANALYTIC GEOMETRY.

It is evident from the construction that SG and SG' can have but one point each in common with the curve; for this reason

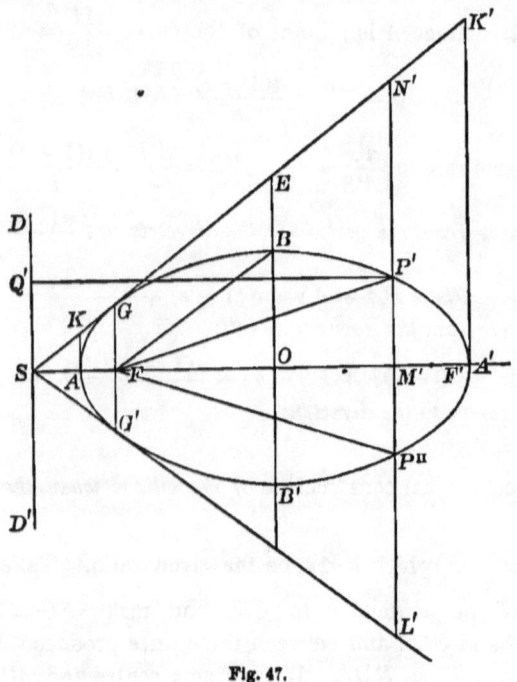

Fig. 47.

they are called the **focal tangents**, and since $GF < FS$, their included angle is less than 90°. So long as $e = \dfrac{k}{s}$ remains the same, the distance SF, taken to represent s, simply determines the scale to which the ellipse is constructed, the angle between the focal tangents remaining the same. But if e varies, the angle GSG' will vary, and the ellipse will change in shape as well as in size.

COR. 1. Since A is a point of the curve $\dfrac{AF}{AS} = e$. But

$$\frac{AK}{AS} = e, \therefore AF = AK.$$

Similarly $A'F = A'K'$. Hence, *to find the focal tangents when the axes are given*, first determine the focus (Art. 58) F; then make $AK = AF$ and $A'K' = A'F$; KK' will be the focal tangent, and its intersection with the axis produced (S) a point of the directrix.

Cor. 2. From Geometry,

$$OE = \tfrac{1}{2}(AK + A'K') = \tfrac{1}{2}(AF + FA') = a.$$

Hence $FB = OE = a$, as already shown.

Cor. 3. Since the curve is symmetrical with respect to its axes, and $OF = OF'$, there is another directrix on the right of the centre at the same distance from it as DD'.

61. *The circle is a particular case of the ellipse.*

Making $a = b$ in the equation of the ellipse $a^2y^2 + b^2x^2 = a^2b^2$, we have $y^2 + x^2 = a^2$, which is the central equation of the circle whose radius is a (Eq. (2), Art. 47).

Cor. 1. When $a = b$, $e = \dfrac{\sqrt{a^2 - b^2}}{a} = 0$; hence *the eccentricity of the circle is zero.*

Cor. 2. When $a = b$, $c^2 = a^2 - b^2 = 0$; hence *the foci of the circle are at the centre.*

Cor. 3. Since, for the circle, $e = 0$, $\dfrac{a}{e} =$ the distance from the centre to the directrix $= \infty$; hence *the directrix of the circle is at infinity and the focal tangents parallel.*

62. Varieties of the ellipse.

Every equation of the form

$$Ay^2 + Cx^2 + F = 0, \qquad (1)$$

which is not impossible, is the central equation of an ellipse, if A and C have like signs and neither is zero.

First. Let F be negative. Then $Ay^2 + Cx^2 = F$. If this is the central equation of an ellipse, it is reducible to the form

$a^2y^2 + b^2x^2 = a^2b^2$, in which the absolute term is the product of the coefficients of the squares of x and y. Let R be the factor which renders $RA \cdot RC = RF$. Then $R = \dfrac{F}{AC}$. Introducing this factor, (1) becomes $\dfrac{F}{C}y^2 + \dfrac{F}{A}x^2 = \dfrac{F^2}{AC}$, which is the required form, and in which, therefore, $a = \sqrt{\dfrac{F}{C}}$, $b = \sqrt{\dfrac{F}{A}}$ are the semi-axes. If $A = C$, the axes become equal, and the ellipse becomes a circle, which we have seen is a particular case of an ellipse.

SECOND. If $F = 0$, (1) becomes $Ay^2 + Cx^2 = 0$, which is true only for $x = 0$, $y = 0$, and the ellipse becomes a point, which, as the limiting case of a circle, may also be considered a variety of the ellipse.

THIRD. If F is positive, (1) becomes $Ay^2 + Cx^2 = -F$, which is impossible, as the first number is the sum of two squares. In this case $y = \sqrt{\dfrac{-F - Cx^2}{A}}$, which is imaginary for all values of x, as are also the semi-axes $\sqrt{\dfrac{-F}{C}}$ and $\sqrt{\dfrac{-F}{A}}$.

There are then four varieties of the ellipse, in which the axes are real and unequal, real and equal, zero, or imaginary; and we may say that *every equation of the form $Ay^2 + Cx^2 + F = 0$, in which A and C have like signs, is the equation of an ellipse, real or imaginary, according as F is negative or positive.*

EXAMPLES. 1. What are the axes of the ellipse $9y^2 + 6x^2 = 20$?
Multiplying by the factor $R = \dfrac{F}{AC} = \dfrac{10}{27}$, the equation becomes
$\tfrac{10}{3}y^2 + \tfrac{20}{9}x^2 = \tfrac{200}{27}$, which is in the form $a^2y^2 + b^2x^2 = a^2b^2$,
and we see by inspection that the axes are $2\sqrt{\tfrac{10}{3}}$ and $2\sqrt{\tfrac{20}{9}}$. Otherwise, since the equation is the central equation, the semi-axes are the intercepts, and we may find them directly by making $x = 0$, $\therefore y = b = \sqrt{\tfrac{20}{9}}$, and $y = 0$, $\therefore x = a = \sqrt{\tfrac{10}{3}}$.

2. Find the axes, eccentricity, and parameter of $3y^2 + 2x^2 = 18$.

Ans. $a = 3$; $b = \sqrt{6}$; $e = \dfrac{1}{\sqrt{3}}$; $2p = 4$.

COMMON EQUATIONS OF CONIC SECTIONS.

3. Find the axes, focal distances, and distance of the directrix from the centre of $6y^2 + 3x^2 = 108$.
    Ans. $a = 6$; $b = 3\sqrt{2}$; $3\sqrt{2}(\sqrt{2} \pm 1)$; $6\sqrt{2}$.

4. Write the equation of the ellipse whose axes are 18 and 10.
    Ans. $81y^2 + 25x^2 = 2025$.

5. Write the equation of the ellipse whose eccentricity is $\frac{2}{3}$ and transverse axis 10.   Ans. $9y^2 + 5x^2 = 125$.

6. The conjugate axis of an ellipse is 4 and its lesser focal distance 1. Find its eccentricity and equation.
    Ans. $\frac{3}{5}$; $25y^2 + 16x^2 = 100$.

7. Construct geometrically the ellipse in the following cases:
    (a) $e = \frac{2}{3}$. Observe the size is undetermined.
    (b) $e = \frac{4}{5}$; distance from focus to directrix $= 10$.
    (c) $a = 5$, $b = 4$.

8. The eccentricity of an ellipse being $\dfrac{1}{\sqrt{3}}$, what is the angle between the focal tangents?   Ans. 60°.

9. Find the focal distances, conjugate axis, parameter, and position of the directrix, of the ellipse $r = \dfrac{28}{16 - 9\cos\theta}$.
    Ans. 4, $\frac{28}{23}$; $\frac{8}{5}\sqrt{7}$; $\frac{7}{2}$; $4\frac{12}{23}$ from centre.

10. Write the polar equation of the ellipse whose axes are 18 and 8.   Ans. $r = \dfrac{16}{9 - \sqrt{65}\cos\theta}$.

THE HYPERBOLA.

63. Defs. *The path of a point so moving that the difference of its distances from two fixed points is constant is called an* **hyperbola.** The two fixed points are called the **foci**, the point midway between them the **centre**, and the lines joining any point of the hyperbola with the foci the **focal radii**.

64. Central equation of the hyperbola.

Let F, F', be the foci, O the centre and origin, the axis of X

being coincident with FF', P any point of the hyperbola, $FF' = 2c$, and $2a$ the constant difference. Then $FP - F'P = 2a$.

But $FP = \sqrt{FM^2 + MP^2} = \sqrt{(x+c)^2 + y^2}$,

$F'P = \sqrt{F'M^2 + MP^2} = \sqrt{(x-c)^2 + y^2}$.

Hence $\sqrt{(x+c)^2 + y^2} - \sqrt{(x-c)^2 + y^2} = 2a$.

Transposing the second term to the second member, and squaring,
$$(x+c)^2 + y^2 = 4a^2 + 4a\sqrt{(x-c)^2 + y^2} + (x-c)^2 + y^2,$$
or $\qquad cx - a^2 = a\sqrt{(x-c)^2 + y^2}$.

Squaring again,
$$a^2 y^2 + (a^2 - c^2) x^2 = a^2(a^2 - c^2). \qquad (1)$$

DISCUSSION OF THE EQUATION. Since only the squares of the variables enter the equation, *the hyperbola is symmetrical with respect to both axes.* Making $y = 0$, the X-intercepts are $\pm a$. Take $OA = OA' = a$, then $AA' = 2a$, the constant difference, and as the difference between two sides of a triangle is less than the third side, $PF - PF' = 2a < FF' = 2c$, or A and A' are *between the foci.* Making $x = 0$, the Y-intercepts are $\pm \sqrt{a^2 - c^2}$, and are imaginary, as $a < c$; hence the curve does not cross the axis of Y. Solving the equation for y,
$$y = \pm \frac{1}{a}\sqrt{(a^2 - c^2)(a^2 - x^2)},$$
which, since $a^2 - c^2$ is negative, is imaginary for all values of $x < a$ numerically, and *the curve lies wholly without the limits A and A' along X, extending to $\pm \infty$.* Solving for x,
$$x = \pm a\sqrt{1 - \frac{y^2}{a^2 - c^2}},$$
which is real for all values of y since $a^2 - c^2$ is negative, or *the curve has no limits in the direction of Y.* The form of the hyperbola is best observed by the following mechanical construction: take a ruler of any length FI, fixing one extremity at F, and a string $F'PI$, shorter than the ruler by $AA' = 2a$, one of whose

COMMON EQUATIONS OF CONIC SECTIONS.   89

ends is attached to the farther extremity I of the ruler, the other at the focus F'. Press the string against the ruler by a pencil, as at P, keeping the string stretched, the ruler turning

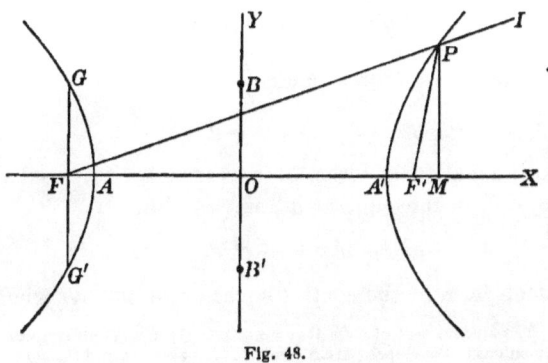

Fig. 48.

about its fixed end F. As the pencil moves it will trace the hyperbola, for in all its positions $FPI = F'PI + 2a$; or, subtracting PI from each number,

$$FP = F'P + 2a, \therefore FP - F'P = 2a.$$

65. Defs. AA' is called the **transverse axis** of the hyperbola, A and A' the **vertices**, FA and FA' (or $F''A'$ and $F''A$) the **focal distances**, the double ordinate through either focus, as GG', the **parameter**, and the distance from the focus to the centre divided by the semi-transverse axis $\left(\dfrac{FO}{AO}\right)$, the **eccentricity**. Equation (1), Art. 64, is called the **central equation** of the hyperbola.

66. Common form *of the central equation.* The hyperbola does not cross the axis of Y, and does not therefore determine by its intercepts a *conjugate* axis, as in the case of the ellipse. Its equation, however, will assume a form similar to that of the ellipse if we represent the *numerical* value of $\sqrt{a^2 - c^2}$ by b, laying off $OB = OB' = b$ (Fig. 48) for a conjugate axis. We thus have $a^2 - c^2 = -b^2$, minus because $a < c$, and $e = \dfrac{c}{a}$, $\therefore c = ae$, e being the eccentricity.

Substituting $c = ae$ in Eq. (1), Art. 64, we have
$$y^2 + (1 - e^2) x^2 = a^2 (1 - e^2),$$
or, since $c > a$, and therefore $e = \dfrac{c}{a} > 1$,
$$y^2 - (e^2 - 1) x^2 = -a^2 (e^2 - 1), \tag{1}$$
the central equation in terms of the eccentricity. Substituting $a^2 - c^2 = -b^2$, in the same equation, we obtain
$$a^2 y^2 - b^2 x^2 = -a^2 b^2, \tag{2}$$
the **common form** of the central equation of the hyperbola.

NOTE. The student will observe that Equations (1) and (2) differ from the corresponding equations of the ellipse (Art. 56) only *in the value of e and the sign of b^2;* also, that while $x = 0$, in Eq. (2), gives $y = \pm \sqrt{-b^2}$, an imaginary quantity (as it should be, since the curve does not cross Y), its *numerical* value is the semi-conjugate axis, as in the case of the ellipse.

Cor. 1. Since $e = \dfrac{c}{a}$, and $c > a$, *the eccentricity of the hyperbola is always greater than unity.*

Cor. 2. Since $a^2 - c^2 = -b^2$, $e = \dfrac{c}{a} = \dfrac{\sqrt{a^2 + b^2}}{a}$, *the eccentricity in terms of the semi-axes.*

Cor. 3. Since $e = \dfrac{c}{a}$, $c = ae = \sqrt{a^2 + b^2} = AB$ (Fig. 48), or *the distance from the focus to the centre is the distance from either vertex to the extremity of the conjugate axis.* Hence, *to find the foci when the axes are given, with O as a centre and AB as a radius, describe an arc; it will cut the transverse axis in the foci.*

67. Length of the focal radii. P being any point of the hyperbola (Fig. 48),
$$FP^2 = FM^2 + MP^2 = (ae + x)^2 + y^2 \text{ (Art. 66, Cor. 3)}$$
$$= (ae + x)^2 + (x^2 - a^2)(e^2 - 1) \text{ (Art. 66, Eq. (1))}$$
$$= e^2 x^2 + 2aex + a^2 = (ex + a)^2;$$
or $\quad FP = ex + a.$

COMMON EQUATIONS OF CONIC SECTIONS.

But $F'P = FP - 2a = ex + a - 2a = ex - a$.

Hence *the focal radii to any point whose abscissa is x are $ex \pm a$.*

68. Polar equation of the hyperbola.

Let the pole be taken at the left-hand focus, and the polar axis coincident with the transverse axis. From the triangle FPF', P being any point of the curve,

$$F'P^2 = FP^2 + FF'^2 - 2 FP \cdot FF' \cos F'FP.$$

But $FP = r$, $F'FP = \theta$, $FF' = 2ae$, and $F'P = FP - 2a = r - 2a$. Making these substitutions, we obtain

$$r = \frac{a(e^2 - 1)}{e \cos \theta - 1}. \tag{1}$$

DISCUSSION OF THE EQUATION. When $\theta = 0°$, $r = a(e+1) = FA'$; when $\theta = 180°$, $r = -a(e-1) = FA$; or *the focal distances are $a(e \pm 1)$, numerically.*

When $\theta = 90°$, $r = -a(e^2 - 1) = -a\left(\frac{c^2}{a^2} - 1\right) = -\frac{b^2}{a}$; or *the parameter, GG', is $2a(e^2 - 1)$, or $\frac{2b^2}{a}$, numerically.*

As θ increases from $0°$, $\cos \theta$ diminishes and r increases, tracing the branch $A'P$, r becoming infinity when $e \cos \theta = 1$,

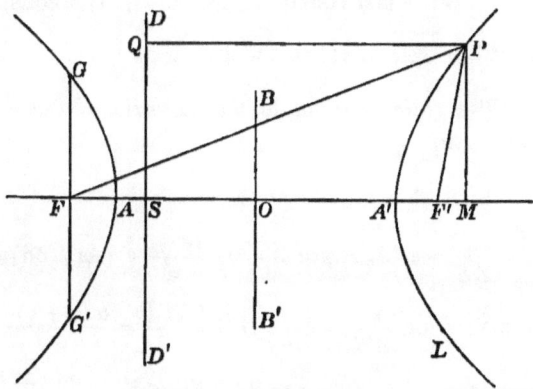

Fig. 49.

or $\theta = \cos^{-1}\dfrac{1}{e}$. When $e\cos\theta > 1$, r is negative, and the branch $G'A$ is traced, in the direction $G'A$, r being FA when $\theta = 180°$. When θ passes $180°$, $\cos\theta$ is negative and r remains negative, tracing the branch AG, and becomes infinity again when $e\cos\theta = 1$, or $\theta = \cos^{-1}\dfrac{1}{e}$ in the fourth angle; after which, $e\cos\theta$ is greater than unity, r is positive and traces the branch LA'.

Representing the parameter $GG' = 2a(e^2-1)$ by $2p$, the polar equation (1) may be written

$$r = \frac{p}{e\cos\theta - 1}. \qquad (2)$$

69. The ratio. *The hyperbola can be traced by a point so moving that the ratio of its distances from a fixed point and a fixed straight line is constant.*

From the polar equation of the hyperbola, $r = \dfrac{p}{e\cos\theta - 1}$, we have $r = er\cos\theta - p$, or (Fig. 49) $FP = eFM - p$. Take FS such that $FS = \dfrac{p}{e}$, or $p = eFS$, and draw DD' perpendicular to FS. Then $FP = e(FM - FS) = eSM = ePQ$, PQ being parallel to MS. But e is a constant; hence $\dfrac{FP}{PQ}$ is a constant.

The fixed line DD' is called the **directrix**.

Cor. 1. *The ratio is equal to the eccentricity, and is always greater than unity.*

Cor. 2. Since A is a point of the curve,

$$\frac{AF}{AS} = e, \therefore AS = \frac{AF}{e} = \frac{a(e-1)}{e} \text{ (Art. 68)}.$$

For the same reason $\dfrac{A'F}{A'S} = e$, $\therefore A'S = \dfrac{A'F}{e} = \dfrac{a(e+1)}{e}$. Hence the distances from the vertices to the directrix are $\dfrac{a(e \mp 1)}{e}$.

Cor. 3. $FS = FA + AS = a(e-1) + \dfrac{a(e-1)}{e} = \dfrac{a(e^2-1)}{e} =$ *the distance from the focus to the directrix.*

Cor. 4. $OS = OF - FS = ae - \dfrac{a(e^2-1)}{e} = \dfrac{a}{e} =$ *the distance from the centre to the directrix.*

70. Geometrical construction *of the hyperbola when the ratio is given.*

Let $e = \dfrac{k}{s}$, in which $k > s$, be the given ratio. Take $FS = s$, draw GG' perpendicular to FS, and make $FG = FG' = k$. Draw SG and SG', and between these lines produced draw any parallel to GG', as $N'L'$. With F as a centre and $M'N'$ as a radius, describe an arc, cutting the parallel in P' and P''; these are points of the hyperbola. To prove that P' is a point of the hyperbola, through S draw DD' perpendicular to SF, and $P'Q'$ perpendicular to DD'. Then, from similar triangles,

$$N'M' : M'S :: GF : FS;$$

but $\qquad N'M' = FP'$, $\;M'S = P'Q'$;

hence $\qquad FP' : P'Q' :: GF : FS,$

or $\qquad \dfrac{FP'}{P'Q'} = \dfrac{GF}{FS} = e.$

Since $N_1M_1 > M_1S$ by construction, the arc described with $FP_1 = M_1N_1$, as a radius will determine P_1, P_2, on the right of DD', which may be proved to be points of the hyperbola as above; and in the same manner any number of points may be constructed.

It is evident from the construction that SG and SG' can have but one point each in common with the curve; for this reason they are called the **focal tangents**. Since $GF > FS$, their included angle, is greater than 90°, the distance FS, taken to represent S, simply determines the scale of the construction; but if e varies, the angle $G'SG$ will vary, and the hyperbola will differ in shape as well as size.

94   ANALYTIC GEOMETRY.

Cor. 1. Since A is a point of the curve, $\frac{AF}{AS} = e$. But $\frac{AK}{AS} = e$, $\therefore AF = AK$. Similarly, $A'K' = A'F$. Hence, *to find the focal tangents when the axes are given*, first determine

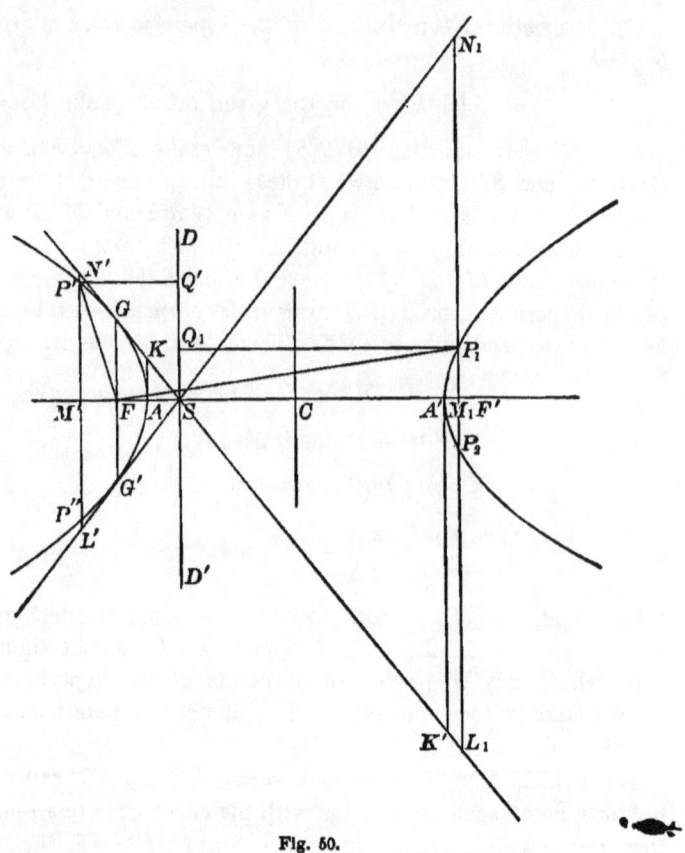

Fig. 50.

the focus (Art. 66, Cor. 3), then make $AK = AF$ and $A'K' = A'F$. KK' will be the focal tangent, and its intersection with the axis, S, a point of the directrix.

Cor. 2. Since the curve is symmetrical with respect to its axes, and $CF'' = CF$, there is another directrix on the right of the centre at the same distance from it as DD'.

71. The equilateral, and the conjugate hyperbola.

When the axes of an hyperbola are equal, it is said to be **equilateral**. Making $a = b$ in the common form of the central equation $a^2y^2 - b^2x^2 = -a^2b^2$, we have

$$y^2 - x^2 = -a^2 \qquad (1)$$

for the equation of the equilateral hyperbola.

Two hyperbolas are said to be **conjugate** *to each other* when the transverse and conjugate axes of the one are the conjugate and transverse axes of the other. If, in deducing the equation of the hyperbola, the transverse axis had been assumed coincident with Y, the equation of the hyperbola in this position would have been $a^2x^2 - b^2y^2 = -a^2b^2$, as this supposition simply amounts to interchanging x and y. Interchanging now a and b, this becomes

$$a^2y^2 - b^2x^2 = a^2b^2; \qquad (2)$$

or, *the central equations of conjugate hyperbolas differ only in the sign of the absolute term.*

Conjugate hyperbolas are distinguished as the X- and the Y-hyperbola, each taking its name from the coordinate axis on which its transverse axis lies, and the equation of either may be derived from that of the other *by changing the signs of a^2 and b^2.*

Cor. 1. The eccentricity of the Y-hyperbola is $\dfrac{\sqrt{a^2+b^2}}{b}$.

Cor. 2. Since the distance of the foci of an hyperbola from the centre is the distance between the extremities of the axes (Art. 66, Cor. 3), *the four foci of a pair of conjugate hyperbolas are equidistant from the centre.*

72. Varieties of the hyperbola.
Every equation of the form

$$Ay^2 + Cx^2 + F = 0 \qquad (1)$$

is the central equation of an hyperbola, if A and C have unlike signs and neither is zero.

FIRST. Let F be positive. Then $Ay^2 - Cx^2 = -F$, which can be reduced to the form $a^2y^2 - b^2x^2 = -a^2b^2$, as in the case of the ellipse (Art. 62), by introducing the factor $R = \dfrac{F}{AC}$; whence $\dfrac{F}{C}y^2 - \dfrac{F}{A}x^2 = -\dfrac{F^2}{AC}$, in which $a = \sqrt{\dfrac{F}{C}}$, and $b = \sqrt{\dfrac{-F}{A}}$ numerically. If $A = C$, the axes are equal and the hyperbola is equilateral.

SECOND. If F is negative, (1) becomes $Ay^2 - Cx^2 = F$, which is the conjugate hyperbola, since it differs from $Ay^2 - Cx^2 = -F$ only in the sign of the absolute term (Art 71).

THIRD. If $F = 0$, (1) becomes $Ay^2 - Cx^2 = 0$, or $y = \pm\sqrt{\dfrac{C}{A}}\,x$, which is the equation of two straight lines through the origin making supplementary angles with X. In this case the axes are zero.

There are then *four varieties of the hyperbola, in which the axes are unequal, equal, interchanged, and zero; corresponding to the X-, equilateral, Y-hyperbola, and a pair of intersecting straight lines through the origin.*

EXAMPLES. 1. What are the axes of the hyperbola

$$9y^2 - 4x^2 = -144?$$

Multiplying by the factor $R = \dfrac{F}{AC} = 4$, the equation becomes

$36y^2 - 16x^2 = -576$, which is of the form $a^2y^2 - b^2x^2 = -a^2b^2$; the axes are therefore 12 and 8. Or, directly, making $y = 0$ and $x = 0$ in succession, we have numerically $x = a = 6$, $y = b = 4$, $\therefore 2a = 12$, $2b = 8$.

2. Find the axes, eccentricity, and parameter of

$$3y^2 - 2x^2 = -18.$$

Ans. $6, 2\sqrt{6}$; $\tfrac{1}{3}\sqrt{15}$; 4.

COMMON EQUATIONS OF CONIC SECTIONS.

3. Find the axes, focal distances, and position of the directrix of $y^2 - x^2 = -81$.
   Ans. $a = b = 9$; $9(\sqrt{2} \pm 1)$; $\dfrac{9}{\sqrt{2}}$ from centre.

4. Write the equation of the hyperbola whose axes are 18 and 10.
   Ans. $81 y^2 - 25 x^2 = -2025$.

5. Write the equation of the hyperbola whose eccentricity is $\frac{4}{3}$ and transverse axis 10.
   Ans. $9 y^2 - 7 x^2 = -175$.

6. The conjugate axis of an hyperbola is 4 and its lesser focal distance 1. Find its eccentricity and write its equation.
   Ans. $\frac{5}{3}$; $9 y^2 - 16 x^2 = -36$.

7. Construct the following hyperbolas:

   (a) $e = \frac{3}{2}$. Observe the size is undetermined.
   (b) $e = \frac{5}{4}$; distance from focus to directrix $= 8$.
   (c) $a = 8$, $b = 6$.

8. Construct a pair of conjugate hyperbolas whose axes are 12 and 8.

9. Write the equations of the hyperbolas conjugate to those of Exs. 2 and 3, and determine their eccentricities and directrices.
   Ans. $\begin{cases} 3 y^2 - 2 x^2 = 18\,; \quad \sqrt{\dfrac{5}{2}}\,; \quad 2\sqrt{\dfrac{3}{5}} \text{ from centre.} \\ y^2 - x^2 = 81\,; \quad \sqrt{2}\,; \quad \dfrac{9}{\sqrt{2}} \text{ from centre.} \end{cases}$

10. The eccentricity of an hyperbola being $\sqrt{3}$, what is the angle between the focal tangents?
    Ans. $120°$.

11. Find the focal distances, conjugate axis, parameter and directrix of the hyperbola $r = \dfrac{6}{\sqrt{15} \cos \theta - 3}$.
    Ans. $\dfrac{6}{\sqrt{15} \mp 3}$; $2\sqrt{6}$; 4; $3\sqrt{\dfrac{3}{5}}$ from centre.

12. Write the polar equation of the hyperbola whose axes are 8 and 6.
    Ans. $r = \dfrac{9}{5 \cos \theta - 4}$.

THE PARABOLA.

73. Defs. *The path of a point so moving that its distance from a fixed point is always equal to its distance from a fixed straight line is called a* **parabola.** The fixed point is the **focus**, the fixed straight line the **directrix**, and the line joining any point of the parabola with the focus, the **focal radius**.

74. Equation of the parabola.

Let F be the focus, DD' the directrix. Draw SF perpendicular to DD' and let $SF = p$. By definition, the middle point O of SF is a point of the curve. Let O be the origin and the axis of X coincident with OF. Then, P being any point of the curve, and PQ perpendicular to DD', $PF = PQ$, or

$$\sqrt{\left(x - \frac{p}{2}\right)^2 + y^2} = x + \frac{p}{2}.$$

Squaring and reducing, $\qquad y^2 = 2px.$ \hfill (1)

DISCUSSION OF THE EQUATION. Solving for y, we have $y = \pm \sqrt{2px}$, or y has two numerically equal and increasing values for positive increasing values of x, but is imaginary when x is negative; hence the curve lies wholly to the right of Y, extends to infinity in the first and fourth angles, and is symmetrical with respect to X. The form of the parabola may be observed from the following mechanical construction: take a ruler of any length QI, and a string, FPI equal in length to the ruler. Fix one end of the string at the focus, the other at the extremity I of the ruler, and, keeping the string pressed against the ruler at P by a pencil, slide the ruler along the directrix parallel to SF; the pencil will trace the curve, for in all its positions $PQ = PF$.

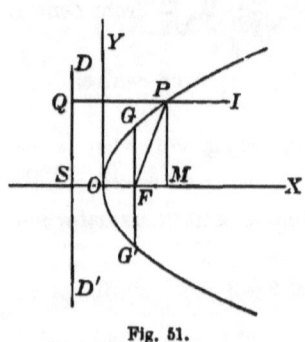

Fig. 51.

COMMON EQUATIONS OF CONIC SECTIONS.

OX is called the **axis** of the parabola, O the **vertex**, and the double ordinate GG' through the focus the **parameter**.

Cor. 1. Substituting $x = OF = \frac{p}{2}$ in Eq. (1), we have $y = FG = p$, or *the parameter $GG' = 2p =$ the coefficient of x in the equation of the curve.* Also $OS = OF = \frac{1}{2}p = \frac{1}{4}GG'$; or $SF = FG = FG' = p$.

Cor. 2. $FP = QP = SO + OM = \frac{1}{2}p + x$; or *the length of the focal radius to any point where abscissa is x is $x + \frac{1}{2}p$.*

75. Polar equation of the parabola.

Let the pole be taken at the focus, and the polar axis coincident with the axis of the parabola. The formulæ for transformation from rectangular axes at O to the polar system, are

$$x = x_0 + r \cos \theta = \frac{p}{2} + r \cos \theta,$$

$$y = y_0 + r \sin \theta = r \sin \theta.$$

Substituting these values in the equation $y^2 = 2px$, we have,

$$r^2 \sin^2 \theta = 2p\left(\frac{p}{2} + r \cos \theta\right),$$

or $\qquad r^2(1 - \cos^2 \theta) = p^2 + 2pr \cos \theta.$

Transposing,

$$r^2 = r^2 \cos^2 \theta + 2pr \cos \theta + p^2 = (r \cos \theta + p)^2.$$

Extracting the root of each member,

$$r = \frac{p}{1 - \cos \theta}, \text{ or } r = \frac{p}{\text{vers } \theta}. \qquad (1)$$

Let the student discuss the equation.

Observe that the equation

$$r = \frac{p}{1 - e \cos \theta} \qquad (2)$$

is the general polar equation of the ellipse, circle, hyperbola, and parabola, when the pole is at the focus; taking the forms

$r = \dfrac{p}{1 - e\cos\theta}$ for the ellipse (Art. 58), that is, when $e<1$; $r=R$ for the circle (Art. 52), when $e=0$; $r = \dfrac{p}{e\cos\theta - 1}$ for the hyperbola (Art. 68), when $e>1$; and $r = \dfrac{p}{1 - \cos\theta}$ for the parabola, when $e=1$.

76. Geometrical construction *of the parabola, the focus and directrix, or the parameter, being given.*

Lay off $SF = p = \frac{1}{2}$ the parameter, or the given distance between the focus and the directrix. Draw GG' perpendicular to SF, and make $FG = FG' = SF$. Draw SG and SG', and any chord $N'L'$ perpendicular to SF. With F as a centre and $M'N'$ as a radius describe an arc cutting the chord in P' and P''. These are points of the parabola. To prove that P' is a point of the parabola, join P' with F, and draw $P'Q'$ parallel and DD' perpendicular to SF. Then

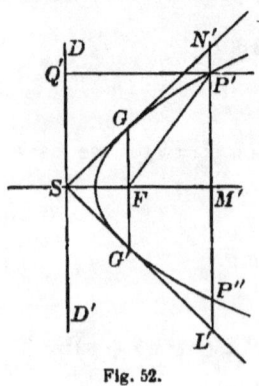

Fig. 52.

$$\dfrac{N'M'}{M'S} = \dfrac{GF}{FS} = \dfrac{P'F}{P'Q'},$$

since the triangles GFS and $N'M'S$ are similar, and

$$P'F = N'M', \; P'Q' = M'S,$$

by construction. In the same way any number of points may be found.

As in the case of the ellipse and the hyperbola, SN' and SL' have evidently but one point each in common with the curve, and are called the **focal tangents**; and as $SF = FG = FG'$, the focal tangents of the parabola make an angle of 90° with each other. $\dfrac{GF}{FS} = \dfrac{P'F}{P'Q'}$ is called the **ratio**, and, evidently, *the ratio of all parabolas is unity.*

COMMON EQUATIONS OF CONIC SECTIONS.

The distance SF, taken to represent p, determines the scale to which the parabola is constructed. Had a distance twice that of the figure been taken, the construction of the same parabola to the new scale would have been equivalent to the construction of a parabola whose parameter was $2\,(2p)$ to the original scale. Hence, *parabolas*, like circles, *differ only in size*.

EXAMPLES. 1. Construct the parabola whose parameter is 10, and write its equation.

2. Construct the parabola the distance of whose vertex from its focus is 2.

3. Write the polar equations of the parabolas of Exs. 1 and 2.

4. The polar equation of a parabola is $r = \dfrac{4}{1 - \cos\theta}$. Write its rectangular equation.

Ans. $y^2 = 8x$.

SECTION VIII. — GENERAL EQUATIONS OF THE CONIC SECTIONS.

77. Defs. *A* **conic** *is the locus of a point so moving that the ratio of its distances from a fixed point and a fixed straight line is constant.* This constant is called the **ratio**, the fixed point the **focus**, the fixed line the **directrix**, and the perpendicular to the directrix through the focus the **axis** of the conic.

78. General equation of the conics.

Let P be any point of the conic, (m, n) the focus F, DD' the directrix, its equation being $x_1 \cos a + y_1 \sin a - p = 0$, the subscripts being used to distinguish the coordinates of the directrix from those of the conic. Then FS, perpendicular to DD', is the axis. Join F with P, draw PQ perpendicular to DD', and let $e =$ the constant ratio. Then $\dfrac{PF}{PQ} = e$, or
$$PF^2 = e^2 PQ^2.$$
But $PF = \sqrt{(y-n)^2 + (x-m)^2}$ (Art. 7); and
$PQ = x \cos a + y \sin a - p$ (Art. 38);

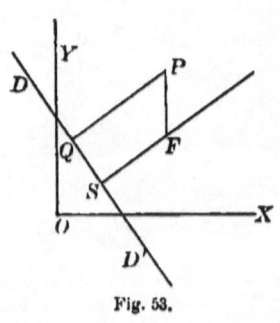

Fig. 53.

hence $\quad (y-n)^2 + (x-m)^2 = e^2 (x \cos a + y \sin a - p)^2 \quad (1)$

is the required equation, in which e determines the species, and m, n, a, and p, the position of the conic.

EXAMPLES. 1. Write the equation of an ellipse whose centre is $(1, 2)$, transverse axis is 6, eccentricity $\dfrac{\sqrt{5}}{3}$, and transverse axis parallel to X.

GENERAL EQUATIONS OF CONIC SECTIONS.

$\alpha = 180°$, $e = \dfrac{\sqrt{5}}{3}$; $\therefore \cos\alpha = -1$, $\sin\alpha = 0$, $p = \dfrac{9}{\sqrt{5}} - 1$, $m = 1 - \sqrt{5}$, $n = 2$.

Substituting in Eq. (1),

$$(y-2)^2 + (x-1+\sqrt{5})^2 = \dfrac{5}{9}\left(-x - \dfrac{9}{\sqrt{5}} + 1\right)^2.$$

or $\quad 9y^2 + 4x^2 - 36y - 8x + 4 = 0.$

2. Write the equation of a parabola whose axis is parallel to X, vertex is at $(-3, -2)$, and parameter is 9.

$\qquad\qquad\qquad\qquad$ Ans. $y^2 + 4y - 9x - 23 = 0$.

3. Write the equation of an ellipse whose eccentricity is $\dfrac{1}{\sqrt{3}}$, centre is $(1,1)$, transverse axis $2\sqrt{3}$, the latter being inclined at an angle $135°$ with X.

$$m = 1 - \dfrac{1}{\sqrt{2}}, \quad n = 1 + \dfrac{1}{\sqrt{2}}, \quad p = 3, \quad e = \dfrac{1}{\sqrt{3}}, \quad \alpha = 135°,$$

and the equation is

$$5y^2 + 2xy + 5x^2 - 12y - 12x = 0.$$

4. Write the equation of a circle whose radius is 5, the axes being tangent to the circle.

$$m = n = 5; \quad y^2 + x^2 - 10y - 10x + 25 = 0.$$

5. The centre of an ellipse is $(-\tfrac{8}{3}, 4)$, its eccentricity $\tfrac{2}{3}$, and its transverse axis $= \tfrac{12}{5}$, and is parallel to X; write its equation.

$\qquad\qquad\qquad$ Ans. $9y^2 + 5x^2 - 72y + 12x + 144 = 0$.

79. *Every complete equation of the second degree between x and y, and all its forms, is the equation of a conic; and, conversely, the equation of every conic is some form of the equation of the second degree.*

Expanding the general equation of the conics, Art. 78, we have

$$\left.\begin{array}{l}(1 - e^2\sin^2\alpha)y^2 - 2e^2\sin\alpha\cos\alpha\, xy + (1 - e^2\cos^2\alpha)x^2 \\ + (2e^2 p\sin\alpha - 2n)y + (2e^2 p\cos\alpha - 2m)x + m^2 + n^2 - e^2 p^2 = 0.\end{array}\right\} (1)$$

The complete equation of the second degree between x and y,

$$Ay^2 + Bxy + Cx^2 + Dy + Ex + F = 0, \qquad (2)$$

is of the same form, but the coefficients of corresponding terms are not necessarily the same, since any equation may be multiplied or divided by any factor without affecting the quality. Making these coefficients, therefore, equal, by dividing each equation by its absolute term, and designating the resulting coefficients of (2) by A', B', etc., we have

$$\left.\begin{array}{r}\dfrac{1-e^2\sin^2 a}{m^2+n^2-e^2 p^2}=A',\\[4pt]\dfrac{-2e^2\sin a\cos a}{m^2+n^2-e^2 p^2}=B',\\[4pt]\dfrac{1-e^2\cos^2 a}{m^2+n^2-e^2 p^2}=C',\\[4pt]\dfrac{2e^2 p\sin a-2n}{m^2+n^2-e^2 p^2}=D',\\[4pt]\dfrac{2e^2 p\cos a-2m}{m^2+n^2-e^2 p^2}=E'.\end{array}\right\} \quad (3).$$

From these five equations the values of the five constants A', B', C', etc., may always be determined when a, m, n, p, and e are given; and as the latter are arbitrary, such values may be assigned to them, that is, the locus may be assumed of such species and in such position, as to give A', B', C', etc., any and every possible set of values. Conversely, the values of a, m, n, p, and e, can always be found from the above equations when those of A', B', C', etc., are given; that is, a conic of some species and position corresponds to any and every set of values which may be assigned to A', B', C', etc. Hence, *every equation of a conic is some one of the forms assumed by the general equation of the second degree, and every form of such equation is the equation of some conic.*

The axes were assumed rectangular. Had they been oblique, the distance FP would have been (Art. 7)

$$\sqrt{(y-n)^2+(x-m)^2+2(y-n)(x-m)\cos\beta},$$

and the distance PQ would have been (Art. 38)

$$x\cos a+y\cos\beta'-p,$$

in which β is the given inclination of the axes, and β' the angle made by PQ with Y. The equation $PF^2 = e^2 PQ^2$ would, therefore, have involved the same arbitrary constants, and no others. Passing now to rectangular axes, since this transformation involves no new arbitrary constants, and cannot affect the degree of the equation, therefore the above reasoning is entirely general.

80. *To determine the species of a conic from its equation.*

Forming $B'^2 - 4A'C'$ from Eq. (3), Art. 79, we have

$$B'^2 - 4A'C'$$
$$= \frac{4e^4 \sin^2 a \cos^2 a - 4(1 - e^2 \sin^2 a)(1 - e^2 \cos^2 a)}{(m^2 + n^2 - e^2 p^2)^2}$$
$$= \frac{4e^4 \sin^2 a \cos^2 a - 4 + 4e^2 \cos^2 a + 4e^2 \sin^2 a - 4e^4 \sin^2 a \cos^2 a}{(m^2 + n^2 - e^2 p^2)^2}$$
$$= \frac{4(e^2 - 1)}{(m^2 + n^2 - e^2 p^2)^2}.$$

Now the locus will be an ellipse, a parabola, or an hyperbola, according as e is less than, equal to, or greater than, unity. But, since the denominator of the above fraction is a square, and the sign of the fraction is thus that of its numerator, when $e < 1$ the first member is negative, when $e = 1$ it is zero, and when $e > 1$ it is positive. Hence the conic will be an ellipse, parabola, or hyperbola, according as $B'^2 - 4A'C'$ is negative, zero, or positive.

To apply this test it is unnecessary to reduce the given equation to the form (1),

$$A'y^2 + B'xy + C'x^2 + D'y + E'x + 1 = 0;$$

for if $B'^2 - 4A'C'$ be negative, zero, or positive, then will

$$(KB')^2 - 4(KA')(KC') = K^2(B'^2 - 4A'C')$$

also be negative, zero, or positive. Hence, *whatever the coefficients, $Ay^2 + Bxy + Cx^2 + Dy + Ex + F = 0$ is the equation of an ellipse, parabola, or hyperbola, according as $B^2 - 4AC$ is negative, zero, or positive.*

EXAMPLES. Determine the species of the following conics:

(1) $y^2 - 5xy + 6x^2 - 14x + 5y + 4 = 0$, *an hyperbola;*

(2) $y^2 - 8xy + 25x^2 + 6y - 2x + 49 = 0$, *an ellipse;*

(3) $3y^2 + 6xy + 4x^2 - 8y = 0$, *an ellipse;*

(4) $y^2 + 2xy + x^2 - y + 1 = 0$, *a parabola;*
(5) $y^2 - 1 + 3x = (x - y)^2$, *an hyperbola;*
(6) $y^2 = 4(x - 1)$, *a parabola;*
(7) $4xy - 16 = 0$, *an hyperbola.*

81. *The equation*
$$Ay^2 + Cx^2 + Dy + Ey + F = 0$$
represents all species of the conic sections.

The general equation of the conics is
$$Ay^2 + Bxy + Cx^2 + Dy + Ex + F = 0.$$

Passing to any rectangular axes with the same origin by the formulæ (Art. 22, Eq. (8)),
$$x = x_1 \cos\gamma - y_1 \sin\gamma, \qquad y = x_1 \sin\gamma + y_1 \cos\gamma,$$
we have, after omitting the subscripts,
$$A(x^2 \sin^2\gamma + 2xy \sin\gamma \cos\gamma + y^2 \cos^2\gamma)$$
$$+ B(x^2 \cos\gamma \sin\gamma + xy \cos^2\gamma - xy \sin^2\gamma - y^2 \sin\gamma \cos\gamma)$$
$$+ C(x^2 \cos^2\gamma - 2xy \cos\gamma \sin\gamma + y^2 \sin^2\gamma)$$
$$+ \text{other terms not involving } xy.$$

The term containing xy is
$$[2A \sin\gamma \cos\gamma + B(\cos^2\gamma - \sin^2\gamma) - 2C \sin\gamma \cos\gamma] xy,$$
or
$$[(A - C) \sin 2\gamma + B \cos 2\gamma] xy,$$
which will be zero if
$$(A - C) \sin 2\gamma + B \cos 2\gamma = 0,$$
or if
$$\tan 2\gamma = \frac{-B}{A - C}. \tag{1}$$

Now $\tan 2\gamma$ can have any and every value from $+\infty$ to $-\infty$, hence a value can always be found for γ which will satisfy (1) whatever the values of A, B, and C; that is, whatever the species of the conic. To find this value of γ, we have (Art. 79)

GENERAL EQUATIONS OF CONIC SECTIONS.

$$\tan 2\gamma = \frac{-B}{A-C} = -\frac{\dfrac{B}{F}}{\dfrac{A-C}{F}} = \frac{-B'}{A'-C'}$$

$$= \frac{2e^2 \sin a \cos a}{1 - e^2 \sin^2 a - (1 - e^2 \cos^2 a)} = \tan 2a,$$

and since $\tan 2a = \tan(180° + 2a)$, (1) will be satisfied whenever $2\gamma = 2a$ or $180° + 2a$; that is, when $\gamma = a$ or $90° + a$, or *whenever the axis of the conic is parallel to either axis of reference.* Hence *every equation of the form*

$$Ay^2 + Cx^2 + Dy + Ex + F = 0$$

is the equation of a conic whose axis is parallel to one of the axes of reference, and, since, $B^2 - 4AC = -4AC$ when $B = 0$, *the conic will be an ellipse, hyperbola, or parabola, according as A and C have like signs, unlike signs, or either is zero* (Art. 80). Thus, whatever the signs or values of D, E, and F,

$$Ay^2 + Cx^2 + Dy + Ex + F = 0 \qquad (2)$$

represents an ellipse whose axes are parallel to the axes of reference;

$$Ay^2 - Cx^2 + Dy + Ex + F = 0 \qquad (3)$$

represents an hyperbola whose axes are parallel to the axes of reference;

or
$$\left. \begin{array}{l} Ay^2 + Dy + Ex + F = 0, \\ Cx^2 + Dy + Ex + F = 0. \end{array} \right\} \qquad (4)$$

represents a parabola whose axis is parallel to X, or Y, respectively.

COR. 1. When referred to the new axes the coefficients of the square are

$$A(\cos^2\gamma + \sin^2\gamma) = A, \quad C(\cos^2\gamma + \sin^2\gamma) = C,$$

or *the coefficients of x^2 and y^2 are not changed by this transformation.*

COR. 2. In the circle $B = 0$, and $A = C$ (Art. 49). Hence $\tan 2\gamma = \frac{0}{0}$, or there is always a pair of axes parallel to the axes of reference.

82. Defs. The centre of a circle is a point equally distant from every point of the circle. The point which has been designated the centre of the ellipse, and hyperbola, is not equally distant from every point of these loci, but it possesses a property in common with the centre of the circle, and in virtue of this common property we may define a centre for all three of these loci. *A locus is said to have a centre when there is a point through which if any chord of the locus be drawn the chord is bisected at that point.*

Any chord through the centre is called a **diameter**.

83. *Every locus whose equation is of the form*
$$Ay^2 + Bxy + Cx^2 + F = 0 \qquad (1)$$
has a centre.

For if (1) be satisfied for any values x', y', of the variables, it is also satisfied for the values $-x'$, $-y'$. But the equation of the chord through (x', y') and $(-x', -y')$ is $x'y - y'x = 0$ (Art. 32), which passes through the origin since it has no absolute term. Moreover, the segments of the chord on either side of the origin are equal, since the length of each is $\sqrt{x'^2 + y'^2}$. Hence the locus has a centre, and the centre is the origin.

COR. 1. Every locus whose equation is of the form
$$Ay^2 + Cx^2 + F = 0$$
has a centre, at the origin.

COR. 2. The circle, ellipse, and hyperbola have centres.

84. *The equation*
$$Ay^2 + Cx^2 + F = 0$$
represents all ellipses and hyperbolas.

Resuming the general equation of the conics,
$$Ay^2 + Bxy + Cx^2 + Dy + Ex + F = 0, \qquad (1)$$
pass to parallel axes, the formulæ for transformation being
$$x = x_0 + x_1, \; y = y_0 + y_1;$$

GENERAL EQUATIONS OF CONIC SECTIONS. 109

and, after omitting subscripts, we obtain

$$A(y_0^2+2y_0y+y^2)+B(x_0y_0+x_0y+y_0x+xy)+C(x_0^2+2x_0x+x^2) \\ +D(y_0+y)+E(x_0+x)+F=0. \quad (2)$$

The terms containing x and y are

$$(2Ay_0+Bx_0+D)y, \quad (2Cx_0+By_0+E)x,$$

which will vanish if $2Ay_0+Bx_0+D=0$, and $2Cx_0+By_0+E=0$; that is, solving these equations for x_0 and y_0, if the new origin is taken at the point

$$x_0=\frac{2AE-BD}{B^2-4AC}, \quad y_0=\frac{2CD-BE}{B^2-4AC},$$

which is always possible when B^2-4AC is not zero; that is, when the locus is not a parabola, in which case x_0 and y_0 would be infinity. Hence the terms containing x and y may always be made to vanish if the locus is an ellipse or an hyperbola, and, when referred to the new axes, the equation will assume the form $Ay^2+Bxy+Cx^2+F=0$, from which we see that *the new origin is the centre* (Art. 83). By a second transformation (Art. 81), the equation will finally take the form

$$Ay^2+Cx^2+F=0,$$

the central equation of the ellipse or hyperbola according as A and C have like or unlike signs.

Cor. 1. Since, when $B^2-4AC=0$, x_0 and y_0 are infinity, *the parabola has no centre.*

Cor. 2. Since the above values of x_0 and y_0 are independent of F, *central conics whose equations differ only in their absolute terms are concentric.*

Cor. 3. By examining Eq. (2) we see that the first three terms of the equation are not altered by the transformation.

85. Varieties of the parabola.

We have seen that when $B^2-4AC=0$ the centre is at infinity, and that therefore the terms Dy and Ex cannot be made

to vanish from the general equation when it represents a parabola; also (Art. 81) that the term Bxy will vanish if either axis of reference is assumed parallel to the axis of the parabola, in which case $B^2 - 4AC$ becomes $-4AC$, and *either* A or C must be zero. Making then $B = 0$ and $C = 0$ in the general equation,
$$Ay^2 + Dy + Ex + F = 0 \qquad (1)$$
represents all parabolas. To see if this form can be still further simplified, transform to new parallel axes by the formulæ $x = x_0 + x_1$, $y = y_0 + y_1$, and we have, omitting subscripts,
$$Ay^2 + (D + 2Ay_0)y + Ex + Ay_0^2 + Dy_0 + Ex_0 + F = 0.$$
As the terms containing x and y cannot both be made to vanish, let us see if one of them, as y, and the absolute term can be made to vanish. This requires that
$$D + 2Ay_0 = 0 \text{ and } Ay_0^2 + Dy_0 + Ex_0 + F = 0,$$
or that $y_0 = -\dfrac{D}{2A}$ and $x_0 = \dfrac{D^2 - 4AF}{4AE}$.

The equation then assumes the form $Ay^2 + Ex = 0$, or
$$y^2 = -\frac{E}{A}x,$$
which is the equation of the parabola referred to its vertex and axis (Art. 74), the curve lying to the right or the left of the origin according as E and A have unlike or like signs. Hence the disappearance of the absolute term and that containing y involves a system of reference *whose origin is the vertex*. This transformation is always possible, except in two cases: *First*, when $E = 0$, in which case $x_0 = \infty$. Equation (1) then becomes $Ay^2 + Dy + F = 0$, or
$$y = \frac{-D \pm \sqrt{D^2 - 4AF}}{2A},$$
which represents two straight lines parallel to X, real and different, real and coincident, or both imaginary, according as D^2 is greater than, equal to, or less than $4AF$. These are the particular cases of the parabola, the vertex receding to infinity.

GENERAL EQUATIONS OF CONIC SECTIONS. 111

Second, when $A = 0$, in which case, however, the equation ceases to be one of the second degree.

86. Defs. A diameter of a conic has been defined as a chord through the centre. As the parabola has no centre it would appear that it has no diameters. A set of lines may, however, be drawn to the parabola which possess a property in common with the diameters of the ellipse and hyperbola; and in virtue of this common property we may define a diameter for all three species of the conics.

A diameter of a conic is the locus of the middle points of parallel chords.

87. *To find the locus of the middle points of parallel chords.*

FIRST. *For the ellipse and hyperbola.*

Let
$$Ay^2 + Cx^2 + F = 0, \qquad (1)$$
in which $A = a^2$, $C = \pm b^2$, $F = \mp a^2 b^2$, as the conic is an ellipse or an hyperbola, be the equation of the locus, and
$$y = a'x + b' \qquad (2)$$
that of any chord PQ. Combining (1) and (2) to find the intersections P and Q, we have, after substituting y^2 from (2) in (1),
$$x^2 + \frac{2a'b'A}{a'^2 A + C}x = -\frac{b'^2 A + F}{a'^2 A + C},$$

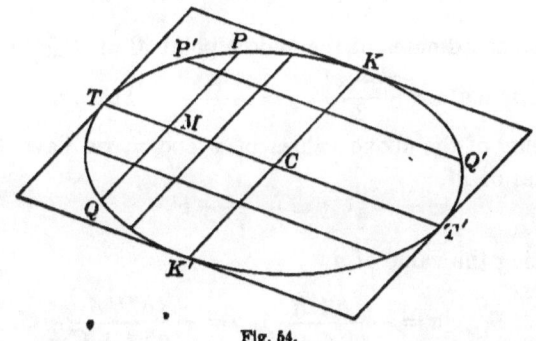

Fig. 54.

or, representing the coefficient of x by q and the absolute term by r,
$$x^2 + qx = r,$$
whence
$$x = -\frac{q}{2} \pm \sqrt{r + \frac{q^2}{4}},$$

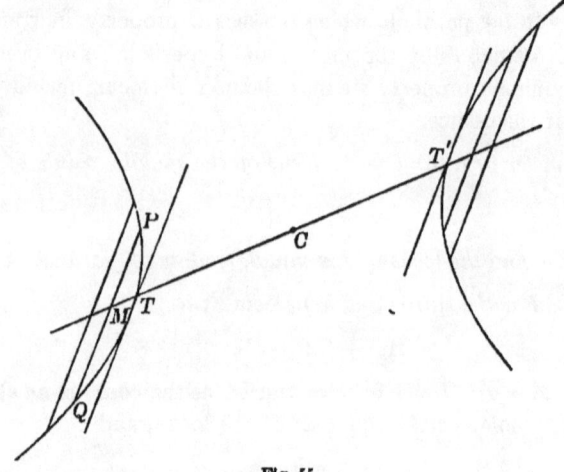

Fig. 55.

which are the abscissas of P and Q. Substituting these values of x in (2), we find the ordinates of P and Q are
$$y = a'\left(\frac{-q}{2} \pm \sqrt{r + \frac{q^2}{4}}\right) + b'.$$

Now the coordinates of the middle point M of PQ are given by the formulæ $x = \frac{x' + x''}{2}$, $y = \frac{y' + y''}{2}$. Taking, therefore, the half-sum of the above values of x and y, we have for the coordinates of M
$$x = -\frac{q}{2}, \quad y = -\frac{a'q}{2} + b',$$
or, replacing the value of q,
$$x = -\frac{a'b'A}{a'^2 A + C}, \quad y = -\frac{a'^2 b' A}{a'^2 A + C} + b'.$$

GENERAL EQUATIONS OF CONIC SECTIONS. 113

For all other chords *parallel to PQ*, a' remains the same, but b' differs. Eliminating then b' by substituting its value from the first in the second of the above equations, we obtain

$$y = -\frac{C}{a'A}x = \mp \frac{b^2}{a'a^2}x, \qquad (3)$$

which is a relation between the coordinates of the middle points of all chords parallel to PQ; it is therefore the equation of a line through these middle points. Being of the first degree it is a straight line, and having no absolute term it passes through the origin, which is the centre. Hence, *the locus of the middle points of parallel chords to the ellipse, or hyperbola, is a straight line through the centre.*

COR. If $A = C$, or the locus is a circle, (3) becomes

$$y = -\frac{1}{a'}x,$$

which is perpendicular to $y = a'x + b'$.

SECOND. *The parabola.*

Let $y^2 = 2px$ be the parabola, and $y = a'x + b'$ any chord PQ. Combining as before,

$$x^2 + \frac{2a'b' - 2p}{a'^2}x = -\frac{b'^2}{a'^2},$$

or, $x^2 + qx = r$, whence, in the same manner the coordinates of the middle point M are

$$x = -\frac{q}{2}, \quad y = -\frac{a'q}{2} + b',$$

or, replacing q by its value,

$$x = \frac{a'b' - p}{a'^2}, \quad y = -\frac{p}{a'},$$

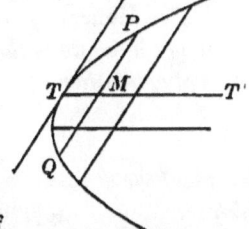

Fig. 56.

from which we see that the abscissa x of the middle point varies with b', but that the ordinate y is constant if a' is constant; that is, if the chords are *parallel*. Hence, *the locus of the middle points of parallel chords to the parabola is a straight line parallel to X.*

114   ANALYTIC GEOMETRY.

The student will observe that if a diameter be defined as a chord through the centre, the diameters of the parabola are necessarily parallel as the centre is infinitely distant.

The extremities of any diameter are called its **vertices.**

88. *The tangents at the vertices of a diameter are parallel to the chords bisected by that diameter.*

Since the diameter TT' (Figs. 54, 55, 56) bisects all chords parallel to PQ, as M approaches T (or T'), P and Q approach each other, and MP, MQ, remaining equal, must vanish together. Hence, when M coincides with T (or T'), PQ will have but one point in common with the curve, or is a tangent.

89. Def. *One diameter is said to be* **conjugate** *to another when it is parallel to the tangents at the vertices of the latter.*

90. Conjugate diameters of the ellipse.

Let KK' (Fig. 54) be drawn parallel to the tangent at T', that is, parallel to PQ. Its equation will be $y = a'x$. The equation of TT' is $y = a''x = -\dfrac{b^2}{a'a^2} x$ (Art. 87, Eq. 3). Hence $a'a'' = -\dfrac{b^2}{a^2}$ is the relation which must exist between the slopes of a diameter and the chords which it bisects. But this relation is satisfied for KK' and the chords $P'Q'$, etc., parallel to TT'. Hence, *if one diameter is conjugate to another, the latter is conjugate to the former, and the tangents at the vertices of conjugate diameters form a parallelogram.*

$$a'a'' = -\frac{b^2}{a^2} \qquad (1)$$

is called *the equation of condition for conjugate diameters to the ellipse.* Since the rectangle of their slopes is negative, the tangents of the angles which they make with X have opposite signs; hence, if one diameter makes an acute angle with the transverse axis, the other will make an obtuse angle, or *conjugate diameters to the ellipse lie on opposite sides of the conjugate axis.*

GENERAL EQUATIONS OF CONIC SECTIONS. 115

Cor. If $a = b$, (1) becomes $a' = -\dfrac{1}{a''}$, or *conjugate diameters to the circle are at right angles to each other.*

91. *Every straight line through the centre of an hyperbola, except the diagonals of the parallelogram on the axes, meets the hyperbola or the conjugate hyperbola.*

Let
$$y = a'x \qquad (1)$$
be any straight line through the centre,
$$a^2 y^2 - b^2 x^2 = -a^2 b^2 \qquad (2)$$
the equation of the X-hyperbola, and (Art. 71)
$$a^2 y^2 - b^2 x^2 = a^2 b^2 \qquad (3)$$
that of the Y-hyperbola. Combining (1) in succession with (2) and (3), we have

$$x^2 = \frac{a^2 b^2}{b^2 - a^2 a'^2}, \quad (4) \qquad x^2 = \frac{a^2 b^2}{a^2 a'^2 - b^2}. \quad (5)$$

Now if $a' < \dfrac{b}{a}$, x is real in (4) and imaginary in (5), and the

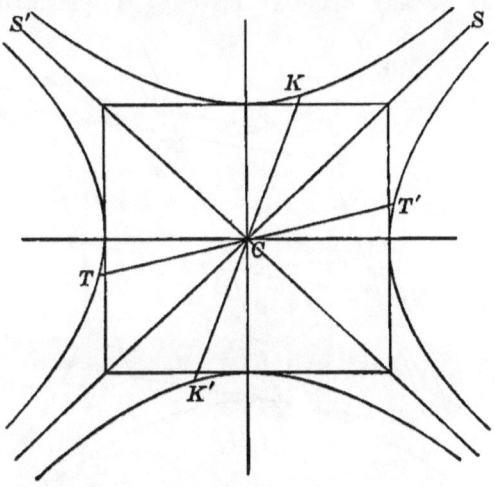

Fig. 57.

line intersects the X-hyperbola, as TT'. If $a' > \frac{b}{a}$, x is imaginary in (4) and real in (5), and the line intersects the Y-hyperbola, as KK'. If $a' = \pm \frac{b}{a}$, both values of x are infinity. In this case (1) becomes $y = \pm \frac{b}{a}x$, the equations of CS and CS', the diagonals of the rectangle on the axes, neither of which meet either hyperbola within a finite distance.

92. Defs. The diagonals of the rectangle on the axes of a pair of conjugate hyperbolas are called the asymptotes. Their equations being $y = \pm \frac{b}{a}x$, if $a = b$ their included angle is 90°, and the hyperbola is said to be **rectangular**; or, *when an hyperbola is rectangular it is also equilateral* (Art. 71).

93. Conjugate diameters of the hyperbola.

Of two conjugate diameters, one meets the X-, the other the Y-hyperbola.

Let TT' be any diameter bisecting a system of parallel

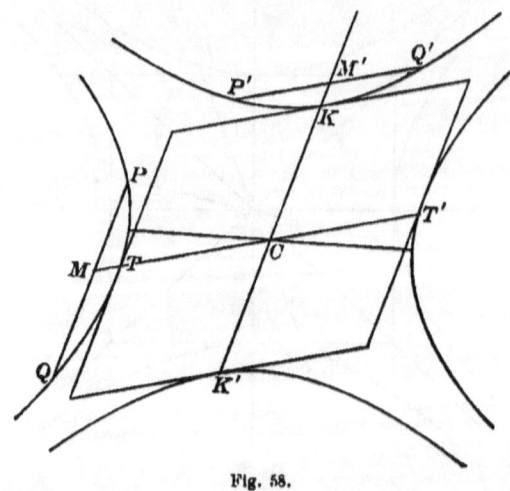

Fig. 58.

GENERAL EQUATIONS OF CONIC SECTIONS. 117

chords of which PQ is one. Draw KK' parallel to PQ, that is, to the tangents at the vertices of TT'; it is then conjugate to TT'. Being parallel to PQ, its equation is $y = a'x$, and that of TT' is $y = a''x = \dfrac{b^2}{a^2 a'}x$ (Eq. 3, Art. 87). Hence $a'a'' = \dfrac{b^2}{a^2}$ is the relation which must exist between the slopes of a diameter and the chords which it bisects. If $a' < \dfrac{b}{a}$, a'' must evidently be $> \dfrac{b}{a}$, since their product $= \dfrac{b^2}{a^2}$, and conversely; or, since $\dfrac{b}{a}$ is the slope of the asymptote, *if one diameter intersects the X-hyperbola, its conjugate will intersect the Y-hyperbola, and conversely.*

Again; since the equation of the Y-hyperbola is derived from that of the X-hyperbola by changing the signs of a^2 and b^2 (Art. 71), $a'a'' = \dfrac{b^2}{a^2}$ is also the relation which must exist between the slopes of any diameter of the Y-hyperbola and the chords which it bisects. But this relation is satisfied for KK' and the chords $P'Q'$, etc., parallel to TT'; hence TT' is parallel to the tangents at K and K', or is conjugate to KK'; hence, *if one diameter is conjugate to another, the latter is conjugate to the former*, and, as in the case of the ellipse, *the tangents at the vertices of conjugate diameters form a parallelogram.*

The equation $$a'a'' = \dfrac{b^2}{a^2}$$
is called *the equation of condition for conjugate diameters to the hyperbola.* Since $a'a''$ is positive, the angles which two conjugate diameters to an hyperbola make with the transverse axis are both acute, or both obtuse, or *the diameters lie on the same side of the conjugate axis.*

CONSTRUCTION OF CONICS FROM THEIR EQUATIONS.

94. FIRST METHOD. *By comparison with the general equation.*

Make the coefficients of like terms in the given and the

general equation equal by dividing each equation by the coefficient of the same term. Equating the resulting coefficients of corresponding terms, we have five equations from which a, m, n, e, and p, may be determined. This method is tedious and of little practical value except as $e = 1$, or some of the coefficients are zero.

EXAMPLE. 1. $y^2 + 4y + 4x + 4 = 0$. Since $B^2 - 4AC = 0$, the conic is a parabola, and therefore $e = 1$. The coefficient of y^2 being unity, divide the general equation (Eq. 1, Art. 79) by the coefficient of y^2, $1 - e^2 \sin^2 a$, and we have, after making $e = 1$,

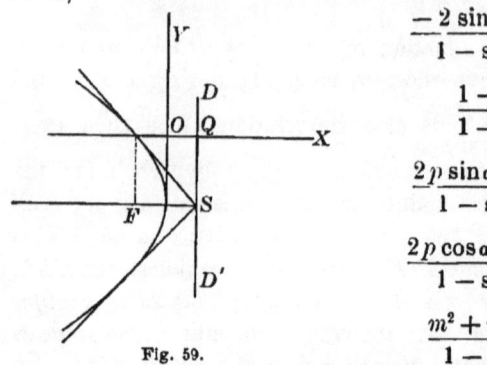

Fig. 59.

$$\frac{-2 \sin a \cos a}{1 - \sin^2 a} = 0, \quad (1)$$

$$\frac{1 - \cos^2 a}{1 - \sin^2 a} = 0, \quad (2)$$

$$\frac{2p \sin a - 2n}{1 - \sin^2 a} = 4, \quad (3)$$

$$\frac{2p \cos a - 2m}{1 - \sin^2 a} = 4, \quad (4)$$

$$\frac{m^2 + n^2 - p^2}{1 - \sin^2 a} = 4. \quad (5)$$

From (1), $-2 \sin a \cos a = 0$; $\therefore a$ must be $0°$, $90°$, $180°$, or $270°$. From (2), $\cos a = \pm 1$; hence a cannot be $90°$ or $270°$, and is either $0°$ or $180°$. In either case (3) gives $n = -2$. Substituting $\cos a = \pm 1$ in (4), we have $\pm 2p - 2m = 4$, or $m = \pm p - 2$, according as a is $0°$ or $180°$. From (5), since $n = -2$, $m^2 = p^2$; or, substituting the above values of m, $p = \pm 1$. But p is always positive; taking, therefore, the upper sign, $a = 0°$. Finally, from (4), making $\cos a = 1$ and $p = 1$, we have $m = -1$. The values of the constants are thus: $e = 1$, $a = 0°$, $m = -1$, $n = -2$, $p = 1$. To construct these results, lay off $OQ = p = 1$ to the right, since $a = 0°$, and draw the directrix DD' perpendicular to X. Construct F, $(-1, -2)$, and through F draw FS perpendicular to DD'.

GENERAL EQUATIONS OF CONIC SECTIONS. 119

Having thus the focus and directrix, the parabola may be constructed as in Art. 76.

95. Second Method. *By transformation of axes.*

If $B^2 - 4AC$ is not zero, the conic is an ellipse or hyperbola, and
$$x_0 = \frac{2AE - BD}{B^2 - 4AC}, \quad y_0 = \frac{2CD - BE}{B^2 - 4AC},$$
are the coordinates of the centre (Art. 84). Transferring to parallel axes with (x_0, y_0) as a new origin, we have the equation of the ellipse, or hyperbola, referred to its centre. If the term Bxy is not present in the primitive equation, the result of this transformation is the central equation of the ellipse, or hyperbola. If, however, this term is present, we must transfer again to new axes with the same origin, the angle between the new and primitive axes of X being determined by the condition $\tan 2\gamma = \dfrac{-B}{A - C}$ (Art. 81). If $B^2 - 4AC = 0$, the conic is a parabola. Transfer first to new axes with the same origin, the new axes of X being subject to the condition $\tan 2\gamma = \dfrac{-B}{A - C}$; then to parallel axes whose origin is (Art. 85)
$$x_0 = \frac{D^2 - 4AF}{4AE}, \quad y_0 = \frac{-D}{2A};$$
the resulting equation will be the equation of the parabola referred to its vertex and axis.

Examples. 1. $5y^2 + 2xy + 5x^2 - 12y - 12x = 0$.
$$B^2 - 4AC = -96,$$
hence the conic is an ellipse and has a centre. The coordinates of the centre are
$$x_0 = \frac{2AE - BD}{B^2 - 4AC} = 1, \quad y_0 = \frac{2CD - BE}{B^2 - 4AC} = 1,$$
and the formulæ of transformation are
$$x = x_0 + x_1 = 1 + x_1, \quad y = y_0 + y_1 = 1 + y_1.$$

Substituting these in the given equation, omitting subscripts, we have
$$5y^2 + 2xy + 5x^2 - 12 = 0.$$
To obtain the central equation we must have
$$\tan 2\gamma = \frac{-B}{A-C} = -\infty \;;\; \therefore \gamma = -45°,$$
and the formulæ of transformation are
$$x = x_1 \cos\gamma - y_1 \sin\gamma = \sqrt{\tfrac{1}{2}}(x_1 + y_1),$$
$$y = x_1 \sin\gamma + y_1 \cos\gamma = \sqrt{\tfrac{1}{2}}(y_1 - x_1).$$
Substituting these values in $5y^2 + 2xy + 5x^2 - 12 = 0$, and omitting subscripts, we obtain $3y^2 + 2x^2 = 6$. The axes are therefore $2\sqrt{3}$ and $2\sqrt{2}$, and the eccentricity $\frac{1}{\sqrt{3}}$.

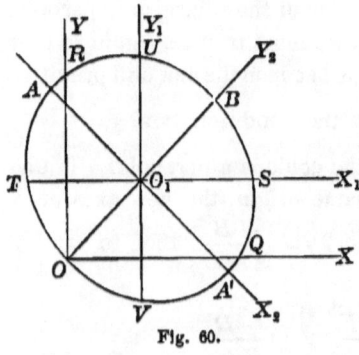

Fig. 60.

To construct the ellipse, construct (1, 1), the new origin O_1, and draw O_1X_1, O_1Y_1, the parallel axes. Draw O_1X_2 making the angle $X_1O_1X_2 = -45°$, and O_1Y_2 perpendicular to it. On O_1X_2 lay off
$$O_1A = O_1A' = \sqrt{3},$$
and on O_1Y_2, $O_1B = O_1O = \sqrt{2}$. AA' and OB are the axes of the ellipse; the focus may be found as in Art. 58, and the ellipse constructed as in Arts. 54 and 60. The curve may be traced with approximate accuracy by determining the intercepts on the axes. Thus, from $5y^2 + 2xy + 5x^2 - 12y - 12x = 0$, $x = 0$ gives $y = 0$ and $\tfrac{12}{5}$ (O and Q); $y = 0$ gives $x = 0$ and $\tfrac{12}{5}$ (O and R). In the same way from $5y^2 + 2xy + 5x^2 - 12 = 0$, $O_1S = O_1T = \sqrt{\tfrac{12}{5}}$, $O_1U = O_1V = \sqrt{\tfrac{12}{5}}$. Through the points thus found trace the curve.

2. $y^2 - 2xy + x^2 + 8x - 16 = 0$. $B^2 - 4AC = 0$, hence the conic is a parabola. $\tan 2\gamma = \frac{-B}{A-C} = \infty$, $\therefore \gamma = 45°$, and the

GENERAL EQUATIONS OF CONIC SECTIONS. 121

formulæ of transformation are $x = \sqrt{\tfrac{1}{2}}\,(x-y)$, $y = \sqrt{\tfrac{1}{2}}\,(x+y)$, and the transformed equation $2y^2 - 4\sqrt{2}\,y + 4\sqrt{2}\,x - 16 = 0$.

From the latter, $x_0 = \dfrac{D^2 - 4AF}{4AE} = \dfrac{5}{\sqrt{2}}$, $y_0 = -\dfrac{D}{2A} = \sqrt{2}$.

Transferring to parallel axes with the origin $\left(\dfrac{5}{\sqrt{2}},\ \sqrt{2}\right)$, we find $y^2 = -2\sqrt{2}\,x$. To construct the parabola, draw the axes $Y_1 O X_1$, making $XOX_1 = 45°$. On these axes construct the vertex $\left(\dfrac{5}{\sqrt{2}},\ \sqrt{2}\right)$, or O_1, and draw the parallel axes $X_2 O_1 Y_2$. We may now construct the parabola whose parameter is $2\sqrt{2}$ as in Arts. 74 or 76, or determine the intercepts and trace the curve approximatively. $OQ = -4 + 4\sqrt{2}$, $OQ' = -4 - 4\sqrt{2}$, $OR = OR' = 4$, $OS = 2\sqrt{2}$, $OT = \sqrt{2} + \sqrt{10}$, $OT' = \sqrt{2} - \sqrt{10}$.

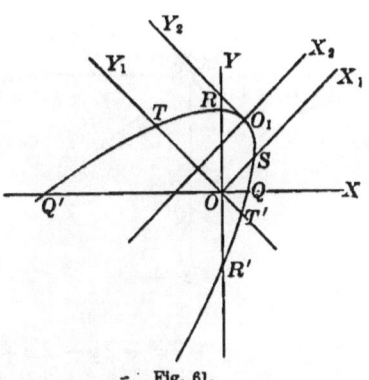

Fig. 61.

3. $y^2 - x^2 + y - x + 2 = 0$. $B^2 - 4AC = 4$, hence the conic is an hyperbola, its axes being parallel to the axes of reference, since $B = 0$. $x_0 = -\tfrac{1}{2}$, $y_0 = -\tfrac{1}{2}$. Passing to parallel axes whose origin is $(-\tfrac{1}{2}, -\tfrac{1}{2})$, we have $y^2 - x^2 = -2$, an equilateral hyperbola whose axes are $2\sqrt{2}$, eccentricity is $\sqrt{2}$, and cutting the primitive axis of X at 1 (Q) and -2 (Q').

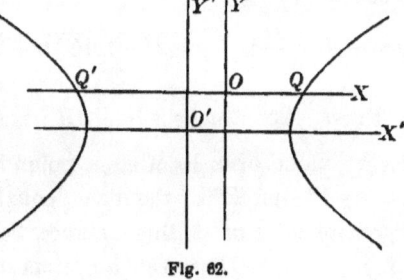

Fig. 62.

4. $y^2 + 2\sqrt{3}xy - x^2 - 64 = 0$. $B^2 - 4AC = 16$, hence the conic is an hyperbola referred to its centre (Art. 83).

$\tan 2\gamma = -\sqrt{3}$, $\therefore 2\gamma = -60°$, or $\gamma = -30°$. Transferring to new axes such that

$$XOX_1 = -30°,$$

the equation becomes

$$y^2 - x^2 = 32,$$

the equilateral Y-hyperbola whose axes are $8\sqrt{2}$, cutting the primitive axis of Y at ± 8 (Q and Q').

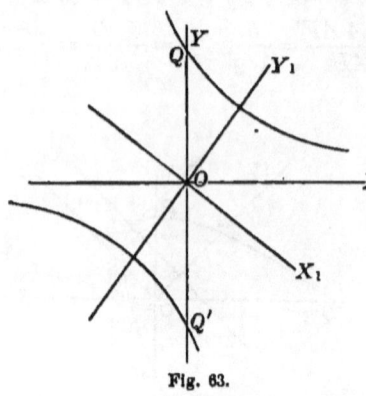

Fig. 63.

5. $y^2 - 4xy + 2x^2 + 2y - 2x + 3 = 0$.
6. $2x^2 + 4xy + 3y^2 - 3 = 0$.

96. Third Method. *By conjugate diameters.*

The general equation of a conic being

$$Ay^2 + Bxy + Cx^2 + Dy + Ex + F = 0,$$

solving for y, we have

$$y^2 + \frac{Bx + D}{A}y = -\frac{Cx^2 + Ex + F}{A},$$

whence

$$y = -\frac{Bx + D}{2A} \pm \frac{1}{2A}\sqrt{(B^2 - 4AC)x^2 + 2(BD - 2AE)x + D^2 - 4AF}.$$

First. Construct the line QR whose equation is $y = -\frac{Bx + D}{2A}$. Every value of x locating a point M on this line locates two points P' and P'' of the locus, equally distant from QR and on opposite sides of it, this distance being the radical in the value of y. Hence QR bisects a system of chords parallel to Y and is a diameter.

GENERAL EQUATIONS OF CONIC SECTIONS.

SECOND. Values of x which render the radical zero give the same values for y for both the locus and the diameter; hence the values of x found from the equation

$$(B^2 - 4AC)x^2 + 2(BD - 2AE)x + D^2 - 4AF = 0$$

determine the points where the conic cuts QR; that is, the

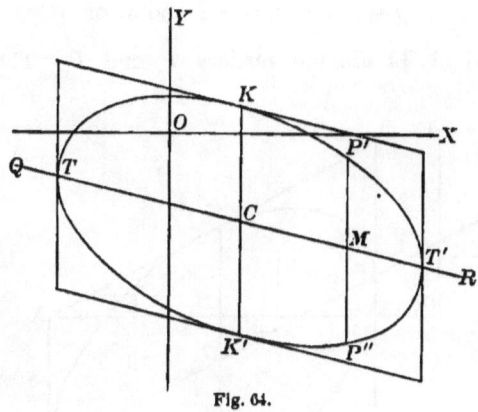

Fig. 64.

vertices T, T', of the diameter. This equation being a quadratic, there will be two such points except when $B^2 - 4AC = 0$, in which case the conic is a parabola and there will be but one vertex. If the conic is an ellipse it lies wholly between T and T'; if an hyperbola, wholly without these points. In either case the half sum of the above values of x determines the centre C, and the corresponding values of y locate K and K', the vertices of the conjugate diameter. Having thus the circumscribing parallelogram (Arts. 90, 93), a few other points may be constructed, especially the intercepts on the axes, and the curve sketched with sufficient accuracy.

EXAMPLES. 1. $4y^2 + 4xy + 5x^2 - 8y - 28x + 24 = 0$.

$$B^2 - 4AC = -64,$$

and the conic is an ellipse. Solving for y we find

$$y = \tfrac{1}{2}(2 - x) \pm \sqrt{-x^2 + 6x - 5}.$$

Construct the diameter
$$y = \tfrac{1}{2}(2-x), \quad QR.$$
Placing
$$-x^2 + 6x - 5 = 0,$$
we find $x = 5$ and 1, whence
$$y = \tfrac{1}{2}(2-x) = -\tfrac{3}{2} \text{ and } \tfrac{1}{2}, \text{ or}$$
$(5, -\tfrac{3}{2})$ and $(1, \tfrac{1}{2})$ are the vertices T' and T. The abscissa

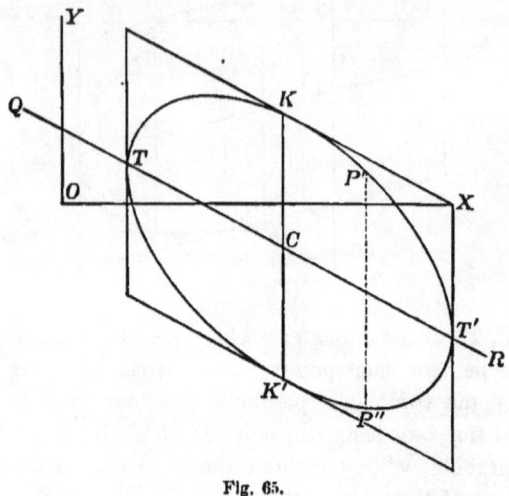

Fig. 65.

of C is $\tfrac{1}{2}(5+1) = 3$, whence, from the equation of the conic $y = \tfrac{3}{2}$ and $-\tfrac{5}{2}$, locating K and K'. The circumscribing parallelogram may now be drawn. Making $y = 0$ we find the X-intercepts $= \dfrac{14 + 2\sqrt{19}}{5}$. Intermediate points may be found if necessary; thus, for $x = 4$, $y = -1 \pm \sqrt{3}$, locating P' and P''. Trace the curve through the points thus found, tangent to the circumscribing parallelogram at K, K', T and T'.

2. $\qquad y^2 + 2xy + x^2 + 2y - 7x - 8 = 0.$

GENERAL EQUATIONS OF CONIC SECTIONS. 125

$B^2 - 4AC = 0$, and the conic is a parabola. Solving for y,

$$y = -(x+1) \pm \sqrt{9x+9}.$$

Construct the diameter QR,

$$y = -(x+1).$$

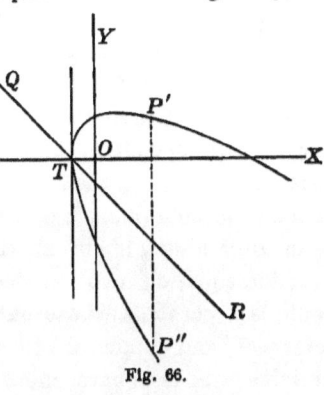

Fig. 66.

The radical gives but one value of $x = -1$, for which $y = 0$, locating the vertex T. The X-intercepts are 8, -1, and the Y-intercepts 2, -4. Intermediate points may also be found; thus for $x = 3$, $y = 2$ and -10 (P' and P''). Trace the curve through these points and tangent at T to a parallel to Y.

3. $y^2 + 2xy - 2x^2 - 4y - x + 10 = 0.$

$B^2 - 4AC = 12,$

∴ the conic is an hyperbola.

$$y = -(x-2) \pm \sqrt{3x^2 - 3x - 6}.$$

The diameter is $y = -(x-2)$, its vertices are $(2, 0)$ and $(-1, 3)$. The X-intercepts are 2, $-\frac{5}{2}$, the Y-intercepts being imaginary.

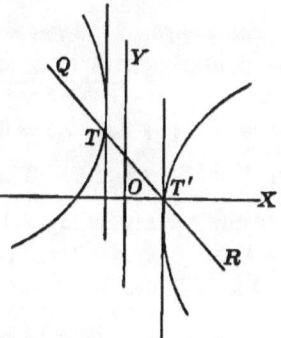

Fig. 67.

4. $y^2 + 2xy + 3x^2 - 4x = 0.$
5. $y^2 - 2xy + x^2 - y + 2x - 1 = 0.$
6. $y^2 - 2xy + x^2 + x = 0.$

97. When the equation of the conic does not contain the term involving xy, the axes of the conic are parallel to the axes of reference, and its position may be determined by the principles of Art. 17. If the squares of both variables are present, it is an ellipse or an hyperbola according as their signs are like or unlike; if these coefficients are equal in magnitude and sign, it is a circle; if numerically equal and of opposite signs, an equilateral hyperbola. Solving the equation for either variable, as y, values of x which render the radical part of y zero give the extremities of the axis parallel to X, and the algebraic difference of these values is the length of this axis; the half sum of these values of x is the abscissa of the centre, and the corresponding values of y determine the vertices of the axis parallel to Y, their algebraic difference being its length. If the term containing x is lacking, the centre is on Y; if the term containing y is absent, the centre is on X.

If the equation involves the square of but one variable, the conic is a parabola whose axis is parallel to the other axis of reference, and coincides with it when the the first power of the variable whose square enters the equation is lacking. The vertex is found by solving the equation for the variable which enters as a square and placing the radical part equal to zero; this equation determines the limit, *i.e.*, the vertex.

EXAMPLES. 1. $9y^2 + 4x^2 - 36y - 8x + 4 = 0$. A and C have like signs, $\therefore$ the conic is an ellipse. Solving for y, $y = 2 \pm \frac{1}{3}\sqrt{-4x^2 + 8x + 32}$. The limits along X are found from $-4x^2 + 8x + 32 = 0$ to be 4 and -2, and the axis parallel to X is therefore 6. The abscissa of the centre is $\frac{4-2}{2} = 1$, and the corresponding values of y are 4, 0, or the axis parallel to Y is 4. Hence the locus is an ellipse whose centre is (1, 2) and axes 6 and 4, its transverse axis being parallel to X.

2. $y^2 + 4y - 6x - 14 = 0$. The locus is a parabola, its axis being parallel to X. Solving for y, $y = -2 \pm \sqrt{6x + 18}$; hence its vertex is $(-3, -2)$.

3. $4y^2 + x^2 + 16y - 4x + 16 = 0$.

4. $9y^2 - 4x^2 - 36y + 24x - 36 = 0$.

GENERAL THEOREMS.

98. *Through any five points in a plane, one conic may be made to pass.*

GENERAL EQUATIONS OF CONIC SECTIONS.

Let (x_1, y_1), (x_2, y_2), (x_3, y_3), (x_4, y_4), (x_5, y_5), be the five given points. Dividing the general equation of a conic by the coefficient of *any* of its terms, and distinguishing the new coefficients by accents, we have

$$A'y^2 + B'xy + C'x^2 + D'y + E'x + 1 = 0. \qquad (1)$$

Substituting in succession the coordinates of the given points, since the conic is to pass through them, we have

$$\left.\begin{aligned}
A'y_1^2 + B'x_1y_1 + C'x_1^2 + D'y_1 + E'x_1 + 1 = 0, \\
A'y_2^2 + B'x_2y_2 + C'x_2^2 + D'y_2 + E'x_2 + 1 = 0, \\
A'y_3^2 + B'x_3y_3 + C'x_3^2 + D'y_3 + E'x_3 + 1 = 0, \\
A'y_4^2 + B'x_4y_4 + C'x_4^2 + D'y_4 + E'x_4 + 1 = 0, \\
A'y_5^2 + B'x_5y_5 + C'x_5^2 + D'y_5 + E'x_5 + 1 = 0,
\end{aligned}\right\} \qquad (2)$$

in which A', B', C', D', and E', are the only unknown quantities, and from which their values may be determined by elimination. Since these equations are of the first degree, each of these quantities has but one value. Substituting in (1) the values of A', B', etc., found from (2), the resulting equation will be that of the conic passing through the five given points.

If one of the points is the origin, one of the equations (2) would be $1 = 0$, which is impossible. In such a case divide the general equation by any coefficient except the absolute term. This results from the fact that the equation sought can have no absolute term.

EXAMPLES. 1. Find the equation of the conic passing through $(4, 4)$, $(4, -4)$, $(9, 6)$, $(9, -6)$, $(0, 0)$.

Since the conic is to pass through the origin, $F = 0$.
Dividing the general equation by A, we have

$$y^2 + B'xy + C'x^2 + D'y + E'x = 0.$$

Substituting the coordinates of the remaining points,

$$16 + 16 B' + 16 C' + 4 D' + 4 E' = 0. \qquad (1)$$
$$16 - 16 B' + 16 C' - 4 D' + 4 E' = 0. \qquad (2)$$
$$36 + 54 B' + 81 C' + 6 D' + 9 E' = 0. \qquad (3)$$
$$36 - 54 B' + 81 C' - 6 D' + 9 E' = 0. \qquad (4)$$

From (1) and (2), and (3) and (4), by addition,

$$32 + 32\,C' + 8\,E' = 0. \qquad (5)$$
$$72 + 102\,C' + 18\,E' = 0. \qquad (6)$$

Eliminating E' between these we find $C' = 0$, which in (5) gives $E' = -4$. Substituting these values in (1) and (3), we have

$$16\,B' + 4\,D' = 0,$$
$$54\,B' + 6\,D' = 0,$$

whence $B' = 0$, $D' = 0$. The required equation is therefore $y^2 = 4x$.

2. Find the equation of the conic passing through $(-1, 2)$, $(-\frac{3}{2}, \frac{1}{2})$, $(-\frac{3}{2}, \frac{7}{2})$, $(-\frac{5}{2}, \frac{1}{2})$, $(-4, 8)$.

$$Ans.\ \ y^2 + x^2 + 2xy + 3x - y + 4 = 0.$$

3. Find the equation of the conic passing through $(5, 0)$, $(0, 5)$, $(-5, 0)$, $(0, -5)$, $(0, 0)$.

In this case $F = 0$, since one of the points is the origin. Divide the general equation by B, otherwise the term Bxy will disappear for every substitution, and B will be undetermined. $Ans.\ The\ Axes.$

4. Through how many points may the conic

$$y^2 + x^2 + Dy + Ex + F = 0$$

be made to pass?

5. Find the equation of the circle circumscribed about the triangle whose vertices are $(3, 1)$, $(2, 3)$, $(1, 2)$.

$$Ans.\ \ 3y^2 + 3x^2 - 11y - 13x + 20 = 0.$$

6. Find the equation of a circle through the origin and making intercepts a and b on the axes.

99. *Two conics can intersect in but four points.*

The coordinates of the points of intersection of two conics will be found by combining their equations (Art. 36). But we know from Algebra that the elimination of one unknown quantity from two quadratic equations gives rise, in general, to an equation of the fourth degree. This equation will have four roots; there will therefore be four sets of coordinates, all four of which may be real, two real and two imaginary (since imaginary roots enter in pairs), or all four imaginary, and in any case, since equal roots occur in pairs, these four sets may reduce to two.

When two sets of values reduce to one, that is, are equal, two of the points of intersection become coincident and the conics are said to *touch each other at that point*. Hence two conics can touch each other at but two points.

The several cases are illustrated in the figure. 1 and 2 intersect in four points, all four sets of values of x and y being real;

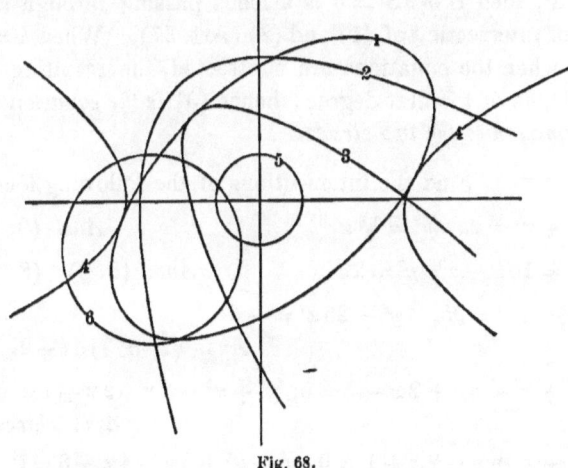

Fig. 68.

1 and 3 intersect each other in two points and touch each other in one, two sets of values being equal; 1 and 4 touch each other in two points; 1 and 5 have no points in common, the roots being all imaginary; while 1 and 6 intersect in two points, two sets of values being real and two imaginary.

If the conics are circles, the simplest way of combining their equations is by subtraction. Thus, let (Art. 48)

$$y^2 + x^2 + Dy + Ex + F = 0, \qquad (1)$$

$$y^2 + x^2 + D'y + E'x + F' = 0, \qquad (2)$$

be the given circles. Subtracting,

$$(D - D')y + (E - E')x + (F - F') = 0, \qquad (3)$$

from which we may find the value of either variable in terms of the other, and substituting it in either (1) or (2) find that of the other.

Cor. 1. Since (3) is of the first degree, two circles can intersect each other in but two points, and hence can touch each other but in one.

Cor. 2. Representing the first members of (1) and (2) by S and S', then $S + kS' = 0$ is a locus passing through all the points of intersection of (1) and (2) (Art. 37). When $k = -1$, that is, when the equations are subtracted, the resulting equation, (3), is of the first degree; hence (3) *is the equation of the chord common to the two circles.*

EXAMPLES. Find the intersections of the following loci:

1. $y^2 + x^2 = 25$, $y^2 = 1\frac{9}{3}x$.   Ans. $(3, \pm 4)$.
2. $y^2 = 10x - x^2$, $y^2 = 2x$.   Ans. $(0, 0)$, $(8, \pm 4)$.
3. $y^2 + 4x^2 = 25$, $4y^2 - 25x^2 = -64$.
     Ans. $(2, \pm 3)$, $(-2, \pm 3)$.
4. $y^2 + x^2 - 3y + 2x - 7 = 0$, $y^2 + x^2 - 3y + 2x + 1 = 0$.
     Ans. Concentric.
5. $y^2 + x^2 + y - 2x + 1 = 0$, $y^2 + x^2 + 3y - 4x + 3 = 0$.
     Ans. $(1, 0)$; $(\frac{1}{2}, -\frac{1}{2})$.
6. $y^2 - 3y + 8x + 10 = 0$, $y^2 + 4x + 6 = 0$.
     Ans. $(-\frac{5}{2}, -2)$; $(-\frac{7}{4}, -1)$.
7. $3y^2 + 2x^2 - 6y + 8x - 10 = 0$, $3y^2 + 2x^2 + 6x - 4 = 0$.
8. $3y^2 + 2x^2 - 6y + 8x - 10 = 0$, $3y^2 + 2x^2 - 6y + 8x + 1 = 0$.
9. Prove that if two circles intersect, the common chord is perpendicular to the line joining their centres.

Let $y^2 + x^2 = R^2$, and $(y - n)^2 + (x - m)^2 = R_1^2$ be the circles. Subtracting these equations, the equation of the common chord is

$$2ny + 2mx - n^2 - m^2 = R^2 - R_1^2, \text{ or } y = -\frac{m}{n}x + C,$$

in which C represents the absolute term. The line through the centres is
$y = \dfrac{n}{m} x$.

10. Prove that the perpendicular from the centre of a circle on a chord bisects the chord.

100. Defs. Conics having the same eccentricity are said to be **similar**. If the corresponding axes are also parallel, each to each, they are said to be similar, and **similarly placed**.

101. *All conics in whose equations the terms of the second degree are the same are both similar and similarly placed.*

Let
$$Ay^2 + Bxy + Cx^2 + D'y + E'x + F' = 0, \qquad (1)$$
$$Ay^2 + Bxy + Cx^2 + D''y + E''x + F'' = 0, \qquad (2)$$

be the equations of the conics, the coefficients of the first three terms being the same. We have seen that $\tan 2\gamma = \dfrac{-B}{A-C}$ (Art. 81), in which $\gamma =$ the angle made by the axis of the conic with X. Since γ depends only upon A, B, and C, the conics are similarly placed.

The axes of the conics being then parallel to each other, we may transform to a system of reference whose axes are parallel to those of both curves; and this transformation does not change the coefficients A and C (Art. 81, Cor. 1), and is possible only when B is the same in both equations. Transforming now each equation separately to an origin at the centre of the corresponding conic and parallel axes, since this transformation does not change the values of A and C (Art. 84, Cor. 3), the equations become

$$Ay^2 + Cx^2 + F_1 = 0, \quad Ay^2 + Cx^2 + F_2 = 0,$$

and the squares of the semi-axes are respectively

$$a'^2 = -\frac{F_1}{C}, \; b'^2 = -\frac{F_1}{A}, \text{ and } a''^2 = -\frac{F_2}{C}, \; b''^2 = -\frac{F_2}{A}.$$

Now $e^2 = 1 \pm \dfrac{b'^2}{a'^2} = 1 \pm \dfrac{b''^2}{a''^2}$, or the eccentricity is the same for each conic; hence they are also similar.

Cor. 1. All parabolas are similar, since $e = 1$ for every parabola.

Cor. 2. All circles are similar and similarly placed.

Cor. 3. If two conics are similar and similarly placed, they can intersect each other in but two points and touch each other in but one. For, subtracting the equations (1) and (2), we have $(D' - D'')y + (E' - E'')x + F'' - F''' = 0$, which is a straight line passing through all the points of intersection of (1) and (2). Combining this equation with (1) or (2), there results two sets of coordinates, which may become equal. The above equation is the equation of the common chord, or tangent.

Cor. 4. If two conics differ only in their absolute terms, they are concentric.

102. *To find the condition that an equation of the second degree may represent two straight lines.*

We have seen that two intersecting straight lines is a particular case of the hyperbola (Art. 72), and that two parallel straight lines is a particular case of the parabola (Art. 85). It is further evident that if we multiply, member by member, two equations of the form $ay + bx + c = 0$, there will result an equation of the second degree, and that this latter will represent the two straight lines represented by the factors, for it will be satisfied by the values of the coordinates which make either of its factors zero. Conversely, *if an equation of the second degree can be resolved into two factors of the first degree, it will represent both the straight lines represented by these factors.* Sometimes these factors can be discovered by simple inspection. Thus, $xy = 0$ can be resolved into the factors $x = 0$, $y = 0$, and, since these are the equations of the axes, $xy = 0$ is the equation of both axes. Again, $x^2 - y^2 = 0$ can be resolved into $x + y = 0$, $x - y = 0$, which are the bisectors of the angles between the axes, and therefore $x^2 - y^2 = 0$ is the equation of both bisectors. As these factors are not always readily seen on inspection of the equation, it becomes desirable to determine the general condition for the existence of two factors of the first degree.

GENERAL EQUATIONS OF CONIC SECTIONS. 133

Let $$Ay^2 + Bxy + Cx^2 + Dy + Ex + F = 0$$
be the general equation of a conic. Solving it for y,
$$y = -\frac{Bx+D}{2A} \pm \frac{1}{2A}\sqrt{(B^2-4AC)x^2+2(BD-2AE)x+D^2-4AF}.$$
In order that this equation may be capable of reduction to the form $y = ax \pm b$, the quantity under the radical must be a perfect square. But the condition that the radical should be a perfect square is
$$(B^2 - 4AC)(D^2 - 4AF) = (BD - 2AE)^2.$$
Expanding and reducing,
$$4ACF + BDE - AE^2 - CD^2 - FB^2 = 0,$$
which is the required condition. If the coefficients of the given equation satisfy this relation, the equation represents two straight lines; to find the lines, solve the equation for y and extract the root indicated by the radical.

EXAMPLES. Determine which of the following equations represent pairs of straight lines, and find the lines:

1. $4y^2 - 5xy + x^2 + 2y + x - 2 = 0.$
Applying the test, we find $-32 - 10 - 4 - 4 + 50 = 0.$
To find the lines, solving for y, we obtain
$$y = \frac{5x-2}{8} \pm \frac{1}{8}\sqrt{9x^2 - 36x + 36} = \frac{5x-2}{8} \pm \frac{1}{8}(3x-6).$$
Hence the lines are $y = x - 1$ and $y = \frac{1}{4}x + \frac{1}{2}.$

2. $3y^2 - 8xy + 3x - 1 = 0.$

3. $y^2 - 2y - x^2 + 1 = 0.$ Ans. $y - x - 1 = 0$, $y + x - 1 = 0.$

4. $xy - ay - bx + ab = 0.$ Ans. $x = a$, $y = b.$

5. $y^2 + 4xy + 4x^2 - 4 = 0.$ Ans. $y + 2x = \pm 2.$

6. $xy + x^2 - y + 9x - 10 = 0.$
   Ans. $x - 1 = 0$, $y + x + 10 = 0.$

SECTION IX. — TANGENTS AND NORMALS.

103. Defs. Let MM' be any locus and AB any secant cutting the locus in the points P' and P''. If AB be turned about P' regarded as fixed, till P'', moving in the locus, coincides with P', AB will then have but one point in common with the locus and is called the **tangent** at that point.

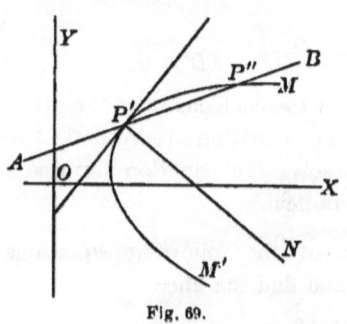

Fig. 69.

The direction of the tangent is that in which the generating point is moving as it passes through the point of tangency, or the slope of the locus at any point is the slope of its tangent at that point. The perpendicular to the tangent at the point of tangency, lying in the plane of the curve, $P'N$, is the **normal**.

104. General equations of the secant and tangent to a conic.

Let
$$y - y' = \frac{y' - y''}{x' - x''}(x - x') \qquad (1)$$

be the equation of a straight line passing through any two given points (x', y'), (x'', y''). The coefficient of x when the equation is solved for y being $\frac{y' - y''}{x' - x''}$, we have $\frac{y' - y''}{x' - x''} = a =$ the slope of the line. Also, let
$$f(x, y) = 0 \qquad (2)$$

be the equation of any conic. If the two points through which the given line passes are on the conic, we must have

TANGENTS AND NORMALS.

$$f(x', y') = 0, \text{ and } f(x'', y'') = 0.$$

Hence, if we form $\dfrac{y' - y''}{x' - x''}$ from these two equations and substitute the value thus found in (1), we shall introduce the condition that (1) is a secant of (2).

Representing this value of $\dfrac{y' - y''}{x' - x''}$ by a_s, the equation of the secant will be

$$y - y' = a_s (x - x').$$

If, now, in the value of a_s, we make $x'' = x'$ and $y'' = y'$, that is, suppose the point (x'', y'') to coincide with (x', y'), the secant will become a tangent; hence, representing what a_s becomes under this supposition by a_t, the equation of the tangent will be

$$y - y' = a_t (x - x'),$$

in which (x', y') is the point of tangency.

EXAMPLES. 1. *Equation of the tangent to the circle*

$$y^2 + x^2 = R^2.$$

Let (x', y'), (x'', y'') be any two points of the circle. Then

$$y'^2 + x'^2 = R^2, \text{ and } y''^2 + x''^2 = R^2.$$

Subtracting, we have

$$y'^2 - y''^2 + x'^2 - x''^2 = 0, \text{ or } (y' - y'')(y' + y'') = -(x' - x'')(x' + x''),$$

whence
$$\dfrac{y' - y''}{x' - x''} = -\dfrac{x' + x''}{y' + y''}.$$

Substituting this value in the equation of a line through two given points,

$$y - y' = \dfrac{y' - y''}{x' - x''}(x - x'),$$

it becomes
$$y - y' = -\dfrac{x' + x''}{y' + y''}(x - x'),$$

the general equation of the secant line to the circle. Making now $x'' = x'$ and $y'' = y'$, we have
$$y - y' = -\dfrac{x'}{y'}(x - x'),$$

for the equation of the tangent. This equation may be simplified by clearing of fractions and replacing $y'^2 + x'^2$ by its equal R^2; whence, finally,

$$yy' + xx' = R^2,$$

the general equation of the tangent to the circle $y^2 + x^2 = R^2$, (x', y') being the point of tangency.

NOTE. The process is the same whatever the equation of the conic; that is, whatever its species or the axes of reference; and the student should thoroughly master the above illustration as exemplifying a *method* for producing the equation of a tangent to any conic when referred to a rectilinear system. Thus, if the circle be referred to a diameter and the tangent at its left-hand vertex, its equation is

$$y^2 = 2Rx - x^2.$$

Hence $\qquad y'^2 = 2Rx' - x'^2 \quad$ and $\quad y''^2 = 2Rx'' - x''^2;$

and, by subtraction, $\quad y'^2 - y''^2 = 2R(x' - x'') - (x'^2 - x''^2);$

whence $\qquad \dfrac{y' - y''}{x' - x''} = \dfrac{2R - (x' + x'')}{y' + y''},$

which becomes $\dfrac{R - x'}{y'}$ when the points coincide. The equation of the tangent is therefore $y - y' = \dfrac{R - x'}{y'}(x - x').$

2. Find the equation of the tangent to the ellipse

$$a^2 y^2 + b^2 x^2 = a^2 b^2. \qquad Ans. \ \ a^2 yy' + b^2 xx' = a^2 b^2.$$

If $a = b$, this becomes $yy' + xx' = a^2$, or R^2, the tangent to the circle, as above. Since the equation of the hyperbola differs from that of the ellipse only in the sign of b^2, we have also the equation of the tangent to the hyperbola, $a^2 yy' - b^2 xx' = -a^2 b^2$.

3. Deduce the equation of the tangent to the hyperbola by the general process.

4. Find the equation of the tangent to the parabola $y^2 = 2px$.

$$Ans. \ \ yy' = p(x + x').$$

When the central equations of the ellipse and hyperbola in terms of the semi-axes are used, and the equation of the parabola referred to its axis and vertex, the corresponding equations of the tangents are easily remembered from the fact that, by dropping the accents which distinguish the coordinates of the point of tangency, they become the equations of the curves themselves.

TANGENTS AND NORMALS.

105. Problems. Under the head of tangency the following simple problems occur.

FIRST. *To write the equation of a tangent at a given point of a conic, and to find the slope of the conic at that point.*

Find the general equation of the tangent to the conic by the preceding method, and substitute in this equation for x', y', the coordinates of the given point. The coefficient of x in the resulting equation, when it is put under the slope form, will be the required slope. Thus, the tangent to the circle $y^2 + x^2 = 100$ at the point $(-6, -8)$ being required, make $x' = -6$, $y' = -8$ in the equation of the tangent $yy' + xx' = R^2$, R being 10, and we have $-8y - 6x = 100$, or $4y + 3x + 50 = 0$, which is the tangent. Solving for y, $y = -\frac{3}{4}x - \frac{50}{4}$, or $a_t = -\frac{3}{4}$, and the angle which the tangent makes with $X = \tan^{-1}\frac{3}{4}$.

SECOND. *To find the point on a conic at which the conic has a given slope.*

In this case the coordinates of the point of tangency are unknown. To find them we have the two equations $f(x', y') = 0$ (since the point is on the conic), and the given condition $a_t = a'$, where a' is the given slope. Combining these equations we find x', y', the required point of tangency. If more than one set of values for x', y', are found, there is more than one solution. If the value of either x' or y' proves to be imaginary, there is no point fulfilling the condition. Thus, at what point of the ellipse $9y^2 + 4x^2 = 36$ does the tangent make an angle of $45°$ with X? By condition, $a_t = -\dfrac{b^2 x'}{a^2 y'} = -\dfrac{4}{9}\dfrac{x'}{y'} = 1$. Also

$$9y'^2 + 4x'^2 = 36.$$

Substituting $x' = -\frac{9}{4}y'$ from the first in the second, we obtain $9y'^2 + 4\frac{81}{16}y'^2 = 36$, or $y' = \pm\sqrt{2}$. Hence $x' = \mp\frac{9}{4}\sqrt{2}$. There are therefore two points at which the slope is 1; one in the second angle, $(-\frac{9}{4}\sqrt{2}, \sqrt{2})$, the other in the fourth,

$$(\tfrac{9}{4}\sqrt{2},\ -\sqrt{2}).$$

THIRD. *To find the equation of a tangent to a conic which passes through a given point without the curve.*

Let $f(x, y) = 0$ be the equation of the conic, and

$$\phi(x, y, x', y') = 0$$

that of the tangent. In this case also the coordinates of the point of tangency are unknown. To find them we have

$f(x', y') = 0$ (since the point of tangency is on the conic),

and $\phi(h, k, x', y') = 0$, in which h, k, are the coordinates of the given point (since the tangent passes through it). Combining these equations, we find x' and y', and there will be as many solutions as there are found sets of values for x', y'. If either x' or y' should prove imaginary, the problem is impossible. Thus, to find the equation of the tangent to the circle $y^2 + x^2 = 25$, passing through $(7, 1)$. Since the point of tangency is on the circle, $y'^2 + x'^2 = 25$. Since the point $(7, 1)$ is on the tangent $yy' + xx' = R^2$, we have also $y' + 7x' = 25$. Combining, we find $x' = 3, y' = 4$, and $x' = 4, y' = -3$. There are therefore two tangents to the circle through $(7, 1)$, namely, $4y + 3x = 25$, and $4x - 3y = 25$.

COR. Since the equation of a conic is of the second degree, and that of the tangent of the first degree, no more than two tangents can be drawn having a given slope, or through a given point without the conic.

EXAMPLES. 1. Find the equation of the tangent to the circle $y^2 + x^2 = 25$ at $(-3, 4)$. *Ans.* $4y - 3x = 25$.

2. Find the slope of the circle $y^2 + x^2 = R^2$ at the points whose abscissas and ordinates are numerically equal.
Ans. $45°$; $135°$.

3. Find the equations of the tangents to the circle $y^2 + x^2 = 100$ passing through the point $(10, 5)$. *Ans.* $4y + 3x = 50$; $x = 10$.

4. Find the slope of the ellipse $3y^2 + x^2 = 3$ at the points $x' = 0$; $y' = 0$; $x' = \frac{3}{2}$. *Ans.* $0°$; $90°$; $135°$ and $225°$.

TANGENTS AND NORMALS.

5. Find the points on the ellipse $8y^2 + 4x^2 = 32$ at which the tangent makes an angle of $135°$ with X.

$$Ans. \left(\frac{4}{\sqrt{3}}, \frac{2}{\sqrt{3}}\right); \left(-\frac{4}{\sqrt{3}}, -\frac{2}{\sqrt{3}}\right).$$

6. Find the tangents to the ellipse $20y^2 + 9x^2 = 324$ passing through $(-1, 6)$.   *Ans.* $5y + 3x = 27$; $35y - 33x = 243$.

7.. Write the equations of tangents to the parabola $y^2 = 8x$ at the points $x' = 8$; $x' = 2$.

8. Find the point on $9y^2 - 4x^2 = -36$ where the tangent makes an angle with X whose tangent is $\frac{5}{3}$.   *Ans. No such point.*

9. Show that the focal tangent to the parabola makes an angle of $45°$ with X. (Find the slope at $y' = p$.)

10. Find the eccentricity of the ellipse $25y^2 + 9x^2 = 225$, by finding the slope at the extremity of the parameter.   *Ans.* $\frac{4}{5}$.

11. Show in the same manner that the eccentricity of the hyperbola $16y^2 - 9x^2 = -144$ is $\frac{5}{4}$.

12. *To write the equation of the tangent to the ellipse in terms of the slope.* The equation of the tangent is $a^2yy' + b^2xx' = a^2b^2$, or $y = -\frac{b^2x'}{a^2y'}x + \frac{b^2}{y'}$. Let $-\frac{b^2x'}{a^2y'} = m =$ slope of the tangent, whose equation then becomes $y = mx + \frac{b^2}{y'}$. To eliminate y', we have

$$b^2x' = -a^2y'm, \quad \text{and} \quad a^2y'^2 + b^2x'^2 = a^2b^2;$$

whence $a^2y'^2 + \frac{a^4y'^2m^2}{b^2} = a^2b^2,$ or $y'^2(a^2m^2 + b^2) = b^4,$

from which we obtain $\frac{b^2}{y'} = \sqrt{a^2m^2 + b^2}$. Thus the equation of the tangent is

$$y = mx + \sqrt{a^2m^2 + b^2}.$$

Changing the sign of b^2, and making $a = b$, we have the corresponding equations for the hyperbola and circle,

$$y = mx + \sqrt{a^2m^2 - b^2}, \quad \text{and} \quad y = mx + a\sqrt{m^2 + 1}.$$

13. *To write the equation of the tangent to the parabola in terms of the slope.*   *Ans.* $y = mx + \dfrac{p}{2m}$.

14. *The rectangle of the perpendiculars from the foci of an ellipse upon the tangent is constant and equal to the square of the semi-conjugate axis.*

Putting the equation of the tangent under the normal form, we have
$$\frac{a^2yy' + b^2xx' - a^2b^2}{\sqrt{a^4y'^2 + b^4x'^2}} = 0.$$

Substituting in succession the coordinates of the foci, ae, 0, and $-ae$, 0, for x and y, and taking the product of the results, we have, putting the radical $= D$ for brevity,

$$\frac{-(b^2aex' - a^2b^2)(b^2aex' + a^2b^2)}{D^2} = \frac{a^4b^4 - b^4a^2e^2x'^2}{D^2}$$
$$= \frac{b^2[a^4y'^2 + a^2b^2x'^2 - b^2a^2e^2x'^2]}{D^2} = \frac{b^2[a^4y'^2 + a^2b^2x'^2(1-e^2)]}{D^2}$$
$$= \frac{b^2(a^4y'^2 + b^4x'^2)}{D^2} = b^2.$$

This property is true also of the hyperbola.

15. *The perpendicular from the focus of an hyperbola upon the asymptote is equal to the semi-conjugate axis.*

This may be regarded as a particular case of the foregoing, the asymptote being a tangent whose point of contact is at an infinite distance. Or, directly, the equation of the asymptote $y = \frac{b}{a}x$ under the normal form is $\frac{ay - bx}{\sqrt{a^2 + b^2}} = 0$; substituting the coordinates of the focus
$$x = ae = \sqrt{a^2 + b^2}, \quad y = 0,$$
this expression reduces to $-b$.

16. *To find the length of a tangent from a given point without a circle.*

Let (x_1, y_1) be the given point P_1, and P' the point of tangency, $(x-m)^2 + (y-n)^2 - R^2 = 0$ being the equation of the circle and C its centre. Then, since the radius to the point of contact is perpendicular to the tangent, $P_1P'^2 = P_1C^2 - CP'^2$. But $P_1C^2 = (x_1 - m)^2 + (y_1 - n)^2$ (Art. 7), $CP'^2 = R^2$. Hence
$$P_1P'^2 = (x_1 - m)^2 + (y_1 - n)^2 - R^2.$$

Now this is what the equation of the circle becomes when the coordinates of the given point are substituted for x and y; hence, put the equation of the given circle under the form $f(x, y) = 0$, and substitute for x and y the coordinates of the given point. The result will be the square of the required distance. Thus the length of the tangent to the circle $y^2 + x^2 - 6y + 8x - 11 = 0$ from (5, 1) is $\sqrt{1 + 25 - 6 + 40 - 11} = 7$.

TANGENTS AND NORMALS.

17. *If two circles are tangent internally and the radius of the larger is the diameter of the smaller, all chords of the larger through the point of contact are bisected by the smaller.*

Take the origin at the point of contact and the diameter as the axis of X. Then the equation of any chord is $y = ax$, and the equations of the larger and smaller circles are $y^2 = 2Rx - x^2$ and $y^2 = Rx - x^2$, respectively. The chord intersects the former at $\left(\dfrac{2R}{a^2+1}, \dfrac{2Ra}{a^2+1}\right)$ and the latter at $\left(\dfrac{R}{a^2+1}, \dfrac{Ra}{a^2+1}\right)$.

106. Chord of contact. *Tangents are drawn to a conic from a given external point; to find the equation of the chord of contact.*

First. *The ellipse and hyperbola.* Let (h, k) be the external point, and (x', y'), (x'', y''), the points of tangency. Then, since both tangents pass through (h, k), these coordinates must satisfy their equations; or

$$a^2 k y' \pm b^2 h x' = \pm a^2 b^2, \qquad (1)$$

$$a^2 k y'' \pm b^2 h x'' = \pm a^2 b^2. \qquad (2)$$

Then $\qquad a^2 k y \pm b^2 h x = \pm a^2 b^2$

is the equation of the chord; for it is satisfied by (x', y'), (x'', y''), as shown by (1) and (2), and is of the first degree with respect to x and y, and therefore represents a straight line through (x', y'), (x'', y'').

Cor. The chord of contact to the circle is $ky + hx = R^2$.

Second. *The parabola.* The equations of both tangents must be satisfied for (h, k); hence

$$ky' = p(h + x'),$$
$$ky'' = p(h + x'').$$

Then $\qquad ky = p(h + x)$

is the chord of contact, the reasoning being identical with that above.

It will be observed that the equations of the chord of contact are derived from those of the tangent by changing the coordinates of the point of tangency, x', y', into those of the external point; these equations are therefore easily memorized.

107. General equation of the normal to a conic. The equation of any line through the point of tangency (x', y') is

$$y - y' = a(x - x').$$

But the normal is perpendicular to the tangent, hence a must equal $-\dfrac{1}{a_t}$, a_t being the coefficient of x in the equation of the tangent when under the slope form; or the equation of the normal is

$$y - y' = -\frac{1}{a_t}(x - x').$$

EXAMPLES. 1. Find the equation of the normal to the circle $y^2 + x^2 = R^2$.

The tangent to the circle is $yy' + xx' = R^2$, or $y = -\dfrac{x'}{y'}x + \dfrac{R^2}{y'}$; hence $a_t = -\dfrac{x'}{y'}$, and the normal is $y - y' = \dfrac{y'}{x'}(x - x')$; or, clearing of fractions, $x'y - y'x = 0$. Since this equation has no absolute term, the normal passes through the origin, which is the centre; hence *the normal to a circle is the radius to the point of tangency.*

2. Find the normal to the ellipse $a^2y^2 + b^2x^2 = a^2b^2$.

$$\text{Ans. } y - y' = \frac{a^2y'}{b^2x'}(x - x').$$

3. Find the normal to the hyperbola $a^2y^2 - b^2x^2 = -a^2b^2$.

$$\text{Ans. } y - y' = -\frac{a^2y'}{b^2x'}(x - x').$$

4. Find the normal to the parabola $y^2 = 2px$.

$$\text{Ans. } y - y' = -\frac{y'}{p}(x - x').$$

5. Find the normal to the circle $y^2 = 2Rx - x^2$.

$$\text{Ans. } y - y' = \frac{y'}{x' - R}(x - x').$$

TANGENTS AND NORMALS. 143

6. Write the equation of a normal in the following cases:
(a) to a circle whose radius is 5 at the point $(3, -4)$.
(b) to an ellipse whose axes are 6 and 4 at the point $x' = 1$.
(c) to a parabola whose parameter is 9 at the point $x' = 4$.
(d) to an hyperbola whose axes are 6 and 4 at the point $x'=8$.

Ans. $\begin{cases} 3y + 4x = 0; \; 3y = \pm 9\sqrt{2}x \mp 5\sqrt{2}; \\ 3y = \mp 4x \pm 34; \; 48y = \mp 9\sqrt{55}x \pm 104\sqrt{55}. \end{cases}$

108. Defs. That portion of the axis of X intercepted between the ordinate from the point of tangency and the tangent is called the **subtangent**. In like manner that portion of the axis of X intercepted between the ordinate and the normal is called the **subnormal**. Thus (Fig. 70), TM and MN are the subtangent and subnormal to the point P'.

109. *To find the subtangent and subnormal at any point of a conic.*

Let $x_t = OT$ represent the intercept of the tangent on the axis of X; that is, the value of x when y is made zero in the equation of the tangent. Then, x' being the abscissa OM of P', the point of tangency,

$TM = \text{subtangent} = OM - OT = x' - x_t$.

Similarly,

$MN = \text{subnormal} = ON - OM = x_n - x'$,

x_n being the X-intercept of the normal, or the value of x when y is made zero in the equation of the normal.

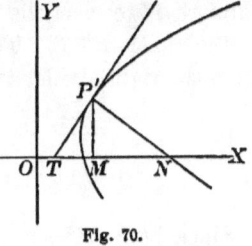

Fig. 70.

EXAMPLES. 1. To find the subtangent of the ellipse and the hyperbola.

The equations of the tangents are $a^2 yy' \pm b^2 xx' = \pm a^2 b^2$. When $y = 0$, $x = x_t = \dfrac{a^2}{x'}$ for both curves. Hence, also, for both curves,

$$\text{subt.} = x' - x_t = \frac{x'^2 - a^2}{x'}.$$

144 ANALYTIC GEOMETRY.

Cor. Since $x_t = \dfrac{a^2}{x'}$, $x_t : a :: a : x'$, or $OT : OA' :: OA' : OM$.

Hence, *the semi-transverse axis is a mean proportional between the intercepts of the tangent and the ordinate of the point of tangency.* (Figs. 71 and 72.)

This principle affords *a method of constructing a tangent at any point.*

First. *The ellipse.* Let P' be the point. Describe the circle $AP'''A'$ on the transverse axis, and produce the ordinate

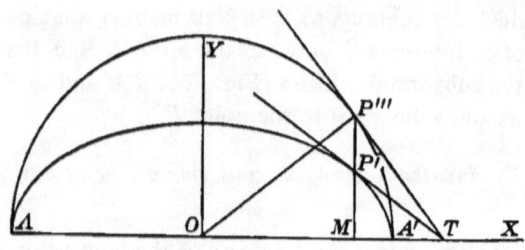

Fig. 71.

through P' to meet the circle at P'''. At P''' draw the tangent to the circle, $P'''T$. Then $P'T$ is the required tangent. For, from the right similar triangles, $OP'''M$, $OP'''T$,

$$OT : OP''' (= OA') :: OP''' : OM;$$

or $\qquad x_t : a :: a : x'.$

Since both $OT = x_t = \dfrac{a^2}{x'}$, and $MT = $ subt. $= \dfrac{x'^2 - a^2}{x'}$, are independent of b, tangents to all ellipses having the same transverse axis, at points having the same abscissa $x' = OM$, will evidently pass through T.

Second. *The hyperbola.* Let P' be the point. Draw the ordinate $P'M$, and on AA', OM, as diameters, describe circles intersecting at Q. Draw QT perpendicular to X. Then TP'

is the required tangent. For, joining Q with O and M, from the similar triangles OQM, OQT, we have

$$OM : OQ(=OA') :: OQ : OT,$$

or $\quad x_t : a :: a : x'.$

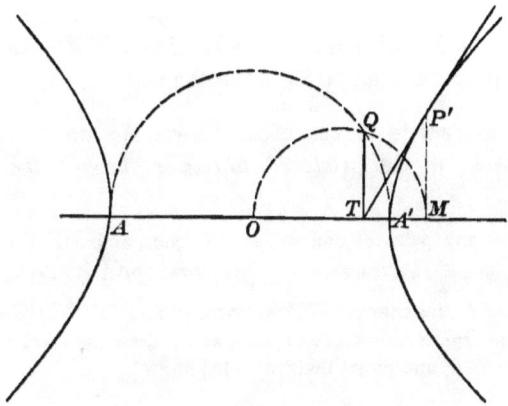

Fig. 72.

Cor. Since $OT = \dfrac{a^2}{x'}$, OT will be zero only when $x' = \infty$; that is, when the tangent coincides with the asymptote (Art. 91).

2. To find the subtangent of the parabola.

Ans. $x_t = -x'$; subt. $= 2x'$.

It appears from this result that *the subtangent of the parabola is bisected at the vertex.* Hence *to construct a tangent at a given point P'*, draw the ordinate $P'M$ and make

$$OT = OM.$$

Then TP' is the required tangent.

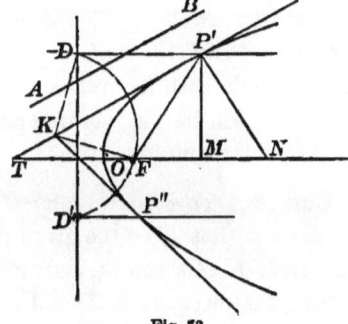

Fig. 73.

3. *To find the subnormal of the parabola.*

The equation of the normal being $y - y' = -\frac{y'}{p}(x - x')$, we have, when $y = 0$, $x = x_n = p + x'$. Hence subn. $= x_n - x' = p$, or *the subnormal of the parabola is constant and equal to one-half the parameter.*

Therefore, *to construct a tangent at a given point,* as P', draw the ordinate $P'M$ and make $MN = p$. Then $P'M$ is the normal, and $P'T$, perpendicular to it, is the tangent.

4. *The tangent to the parabola bisects the angle between the focal radius and the produced diameter through the point of contact.*

Let P' be the point of contact, F the focus, and $P'D$ the diameter. Then $FP' = x' + \frac{p}{2}$ (Art. 74, Cor. 2). Also $TF = TO + OF = x' + \frac{p}{2}$ (Ex. 2). Hence $FT = FP'$, the triangle TFP' is isosceles, and $DP'T = P'TF = FP'T$. Therefore, *to draw a tangent at any point,* as P', draw the focal radius $P'F$, the diameter $P'D$, and bisect their included angle.

Cor. 1. $FN = OM - OF + MN = x' - \frac{p}{2} + p$ (Ex. 3);

$$\therefore FN = x' + \frac{p}{2} = FT = FP',$$

or *the circle described from the focus with a radius equal to the focal radius of any point passes through the intersections of the normal and tangent to that point with the axis.* The triangle $FP'N$ is thus isosceles, and $FNP' = FP'N$.

Cor. 2. $P'FN = FP'T + FTP' = 2\,FTP'$. Hence, *to draw a tangent parallel to a given line,* as AB, from F draw FP' making an angle with the axis equal twice that made by the given line. P' will be the required point of contact and $P'T$, parallel to AB, the required tangent.

Cor. 3. *To draw a tangent through a given point without the curve.* Let K be the given point. Join K with the focus, and with K as a centre and KF as a radius describe a circle cutting the directrix in D and D'. Draw the diameters through D and D'; their intersections with the curve, P' and P'', are

TANGENTS AND NORMALS.   147

the points of tangency. To prove that KP' is a tangent, we have $P'D = P'F$ by definition of the parabola; also $KF = KD$ by construction. Hence KP' bisects the angle $FP'D$. Similarly, $P''K$ may also be shown to be a tangent at P''.

5. *The tangent and normal at any point of the ellipse bisect the angles formed by the focal radii drawn to the point of contact.*

Since the tangent $P'K$ is perpendicular to the normal $P'N$, we have only to prove that $P'N$ bisects $FP'F'$, or that $FN:FP' :: F'N:F'P'$. Now, FP' and $F'P'$ are the focal radii $a + ex'$, $a - ex'$ (Art. 57), respectively.

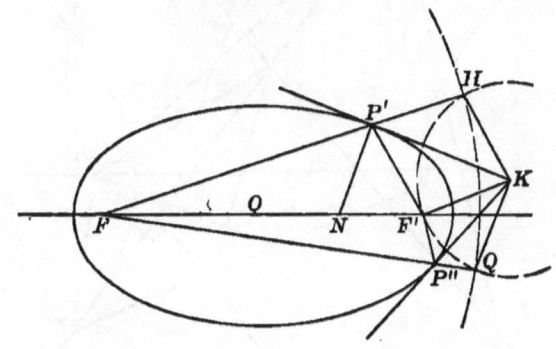

Fig. 74.

$FN = FO + ON$, in which $FO = ae$, and ON is the X-intercept of the normal. Making $y = 0$ in the equation of the normal

$$y - y' = \frac{a^2 y'}{b^2 x'}(x - x'),$$

we have $x = ON = \dfrac{a^2 - b^2}{a^2} x' = e^2 x'.$

Hence $FN = ae + e^2 x' = e(a + ex').$

Also $F'N = F'O - ON = e(a - ex').$

Therefore $\dfrac{FN}{FP'} = \dfrac{e(a + ex')}{a + ex'}$, $\dfrac{F'N}{F'P'} = \dfrac{e(a - ex')}{a - ex'}$, or $\dfrac{FN}{FP'} = \dfrac{F'N}{F'P'}.$

To draw a tangent at any point, as P', we have, obviously, only to bisect the angle FPF' and draw $P'K$ perpendicular to the bisector.

ANALYTIC GEOMETRY.

6. *The tangent and normal at any point of the hyperbola bisect the angles formed by the focal radii drawn to the point of contact.*

Let $P'T$ be the tangent at P'. We have to prove that it bisects the angle $FP'F'$, or that $FT:FP'::F'T:F'P'$. The focal radii FP', $F'P'$, are $ex'-a$ and $ex'+a$ (Art. 67), respectively. $FT=FO-OT$, in which

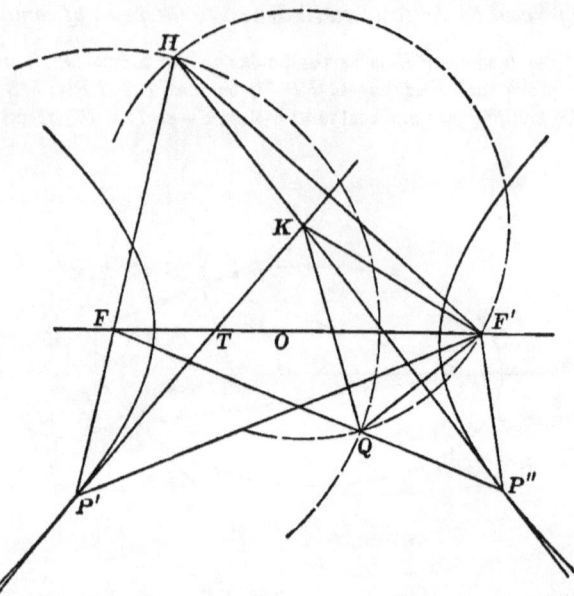

Fig. 75.

$FO = ae$ and OT is the X-intercept of the tangent. Making $y=0$ in the equation of the tangent $a^2yy' - b^2xx' = -a^2b^2$, we have $x = OT = \dfrac{a^2}{x'}$; hence $FT = ae - \dfrac{a^2}{x'} = \dfrac{a}{x'}(ex'-a)$.

Similarly, $F'T = \dfrac{a}{x'}(ex'+a)$, and, as before, $\dfrac{FT}{FP'} = \dfrac{F'T}{F'P'}$.

To draw a tangent at any point, as P', bisect the angle between the focal radii drawn to the point.

7. The principles of Exs. 5 and 6 afford a method for *constructing a tangent passing through a given point without the curves.* Thus let K (Figs. 74, 75) be the given point. Join K

with the *nearer* focus F', and with K as a centre and KF'' as a radius describe an arc. With the *farther* focus F as a centre and FH equal to the transverse axis as a radius describe a second arc cutting the first in H and Q. Join H and Q with the *farther* focus; the intersections P' and P'' of FQ and FH with the curve are the points of tangency. To prove that KP' is a tangent, we have $KH = KF''$, being radii of the same circle; also $P'H = P'F'$, since each is equal to $2a \mp FP'$, the upper sign applying to the ellipse and the lower to the hyperbola. Hence KP' bisects the angle $F'P'H$ in the ellipse and $FP'F'$ in the hyperbola.

Cor. If an ellipse and an hyperbola have the same foci, at the points of intersection they have the same focal radii, and the tangent to the hyperbola is the normal to the ellipse, and conversely. Hence *confocal conics intersect each other at right angles*.

8. Tangents at two points P', P'', *of a parabola, meet the axis in T' and T''. Prove that $T'T'' = FP' - FP''$, F being the focus.*

9. *Two equal parabolas have a common axis but different vertices. Prove that any tangent to the interior, limited by the exterior, parabola, is bisected at the point of contact.*

SECTION X.—OBLIQUE AXES.

CONJUGATE DIAMETERS.

110. Equation of the ellipse referred to conjugate diameters.

Let $A'A'' = 2a'$, $B'B'' = 2b'$, be conjugate diameters, the axes of reference being taken as in the figure. To transform the equation

$$a^2 y^2 + b^2 x^2 = a^2 b^2 \qquad (1)$$

to these axes, we have the formulæ (Art. 22, Eq. 7)

$$x = x_1 \cos \gamma + y_1 \cos \gamma_1, \; y = x_1 \sin \gamma + y_1 \sin \gamma_1.$$

Substituting these in (1), and omitting the subscripts of x and y,

$$\left. \begin{array}{l} (a^2 \sin^2 \gamma_1 + b^2 \cos^2 \gamma_1) y^2 + 2 (a^2 \sin \gamma \sin \gamma_1 + b^2 \cos \gamma \cos \gamma_1) xy \\ + (a^2 \sin^2 \gamma + b^2 \cos^2 \gamma) x^2 = a^2 b^2. \end{array} \right\} (2)$$

But, since the diameters are conjugate, they must fulfil the condition $\tan \gamma \tan \gamma_1 = -\dfrac{b^2}{a^2}$ (Art. 90),

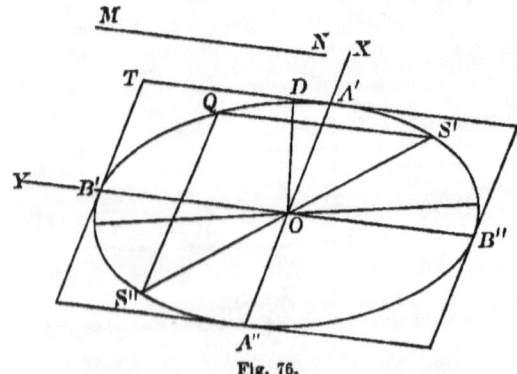

Fig. 76.

or
$$a^2 \sin\gamma \sin\gamma_1 = -b^2 \cos\gamma \cos\gamma_1.$$

Hence the coefficient of the second term of (2) is zero, and the equation becomes

$$(a^2 \sin^2\gamma_1 + b^2 \cos^2\gamma_1) y^2 + (a^2 \sin^2\gamma + b^2 \cos^2\gamma) x^2 = a^2 b^2. \quad (3)$$

Making $y = 0$, we have

$$x^2 = a'^2 = \frac{a^2 b^2}{a^2 \sin^2\gamma + b^2 \cos^2\gamma};$$

and when $x = 0$, $\quad y^2 = b'^2 = \dfrac{a^2 b^2}{a^2 \sin^2\gamma_1 + b^2 \cos^2\gamma_1}.$

Substituting from these equations the values of the coefficients of y^2 and x^2 in (3), we have the equation in terms of the semi-diameters,
$$a'^2 y^2 + b'^2 x^2 = a'^2 b'^2, \quad (4)$$

which is of the same form as the equation of the ellipse referred to its axes, the semi-diameters having replaced the semi-axes.

Cor. 1. The equation of the hyperbola referred to conjugate diameters is
$$a'^2 y^2 - b'^2 x^2 = -a'^2 b'^2, \quad (5)$$

since the only change in the above would be that arising from the minus sign of b^2 in the equation of the hyperbola.

Cor. 2. The equations of the tangents to the ellipse and hyperbola referred to conjugate diameters are

$$a'^2 yy' \pm b'^2 xx' = \pm a'^2 b'^2, \quad (6)$$

since the only change in the process of Art. 104 would be that arising from the substitution of a' and b' for a and b.

111. *The squares of ordinates parallel to any diameter of an ellipse are to each other as the rectangles of the segments into which they divide its conjugate.*

Let $P'M'$, $P''M''$, be the ordinates parallel to any diameter BB', and meeting its conjugate AA' in M' and M''. Then, $a'^2 y^2 + b'^2 x^2 = a'^2 b'^2$ being the equation of the ellipse referred to these diameters, we have for the points P' and P''

$$y'^2 = \frac{b'^2}{a'^2}(a'^2 - x'^2), \quad y''^2 = \frac{b'^2}{a'^2}(a'^2 - x''^2).$$

Dividing, $\quad \dfrac{y'^2}{y''^2} = \dfrac{a'^2 - x'^2}{a'^2 - x''^2} = \dfrac{(a'+x')(a'-x')}{(a'+x'')(a'-x'')},$

or $\qquad P'M'^2 : P''M''^2 :: AM' \cdot M'A' : AM'' \cdot M''A'.$

112. *The squares of ordinates parallel to any diameter of an hyperbola are to each other as the rectangles of the distances from the feet of the ordinates to the vertices of the conjugate diameter.*

113. *The parameter of an ellipse is a third proportional to the transverse and conjugate axes.*

The axes being conjugate diameters, Art. 111 applies, and

$$y'^2 : y''^2 :: (a+x)(a-x') : (a+x'')(a-x'').$$

Let P' coincide with the extremity of the conjugate axis, and P'' with that of the parameter. Then

$$y' = b, \quad y'' = p, \quad x' = 0, \quad x'' = ae,$$

and the proportion becomes

$$b^2 : p^2 :: a^2 : a^2(1-e^2).$$

But $\quad 1 - e^2 = \dfrac{b^2}{a^2}; \quad \therefore a^2 : b^2 :: b^2 : p^2, \quad$ or $\quad 2a : 2b :: 2b : 2p.$

114. *Any ordinate to the transverse axis of an ellipse is to the corresponding ordinate of the circumscribed circle as the conjugate axis of the ellipse is to its transverse axis.*

From the equation of the ellipse $y^2 = \dfrac{b^2}{a^2}(a^2 - x^2)$, and that of the circumscribed circle $y_1^2 = a^2 - x^2$, where y and y_1 are the ordinates corresponding to the same abscissa x, we have

$$\frac{y^2}{y_1^2} = \frac{b^2}{a^2}.$$

115. *The sum of the squares of conjugate diameters to the ellipse is constant and equal to the sum of the squares of the axes.*

OBLIQUE AXES.

Let x', y', be the coordinates of A' (Fig. 76), and x'', y'', those of B'. Since $B''B'$ is parallel to the tangent at A', its equation is $y = -\dfrac{b^2 x'}{a^2 y'} x$. Combining this with $a^2 y^2 + b^2 x^2 = a^2 b^2$, to determine the intersections B' and B'', we find

$$x = x'' = \pm \frac{a y'}{b}, \quad \text{and} \quad y = y'' = \mp \frac{b x'}{a}.$$

But
$$a'^2 = x'^2 + y'^2 = x'^2 + \frac{b^2}{a^2}(a^2 - x'^2) = b^2 + \frac{a^2 - b^2}{a^2} x'^2 = b^2 + e^2 x'^2;$$

and $b'^2 = x''^2 + y''^2 = \dfrac{a^2 y'^2}{b^2} + \dfrac{b^2 x'^2}{a^2} = \dfrac{a^2}{b^2}\left[\dfrac{b^2}{a^2}(a^2 - x'^2)\right] + \dfrac{b^2}{a^2} x'^2$

$$= a^2 + \frac{b^2 - a^2}{a^2} x'^2 = a^2 - e^2 x'^2.$$

Hence $\qquad a'^2 + b'^2 = a^2 + b^2.$

116. *The difference of the squares of conjugate diameters to the hyperbola is constant and equal to the difference of the squares of the axes.*

Let x', y', be the coordinates of A', and x'', y'', those of B'.

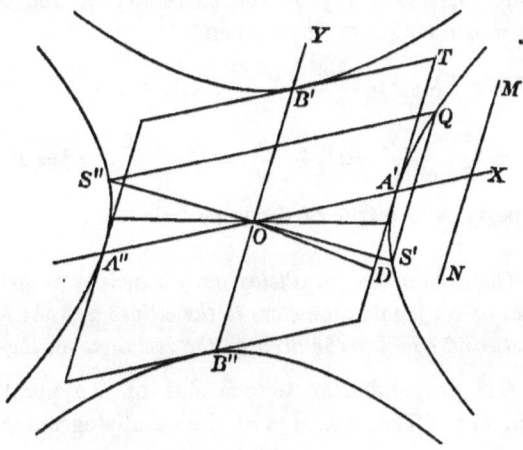

Fig. 77.

The equation of $B''B'$ is $y = \dfrac{b^2 x'}{a^2 y'} x$. Combining this with the equation of the Y-hyperbola, $a^2 y^2 - b^2 x^2 = a^2 b^2$, to determine the intersections B' and B'', we find

$$x = x'' = \pm \frac{ay'}{b}, \quad y = y'' = \pm \frac{bx'}{a}.$$

Hence

$$a'^2 = x'^2 + y'^2 = x'^2 + \frac{b^2}{a^2}(x'^2 - a^2) = \frac{a^2 + b^2}{a^2} x'^2 - b'^2 = e^2 x'^2 - b^2;$$

and $\quad b'^2 = x''^2 + y''^2 = \dfrac{a^2 y'^2}{b^2} + \dfrac{b^2 x'^2}{a^2} = \dfrac{a^2}{b^2}\left[\dfrac{b^2}{a^2}(x'^2 - a^2)\right] + \dfrac{b^2}{a^2} x'^2$

$$= \frac{a^2 + b^2}{a^2} x'^2 - a^2 = e^2 x'^2 - a^2.$$

Hence $\qquad a'^2 - b'^2 = a^2 - b^2.$

117. *The rectangle of the focal radii drawn to the extremities of any diameter of an ellipse is equal to the square of the semi-conjugate diameter.*

Let (x', y') be one extremity of the diameter. Then, if r and r' represent the focal radii, $rr' = (a - ex')(a + ex')$ (Art. 57). Let (x'', y'') be the extremity of the conjugate diameter whose length is $2b'$. Then

$$b'^2 = x''^2 + y''^2 = \frac{a^2 y'^2}{b^2} + \frac{b^2 x'^2}{a^2} \text{ (Art. 115)}$$

$$= \frac{a^2}{b^2} \cdot \frac{b^2}{a^2}(a^2 - x'^2) + \frac{b^2 x'^2}{a^2} = a^2 - \frac{a^2 - b^2}{a^2} x'^2 = a^2 - e^2 x'^2.$$

This property is also true of the hyperbola.

118. *The area of the parallelogram formed by tangents at the extremities of conjugate diameters to the ellipse and the hyperbola is constant, and equal to the area of the rectangle on the axes.*

Draw OD perpendicular to one side of the parallelogram (Figs. 76, 77). Then the area of the parallelogram is

$4\,OB'.\,OA'. \sin B'OA' = 4\,OB'.\,OA'. \sin OA'T = 4\,OB'.\,OD.$

OBLIQUE AXES.

The normal form of the equation of the tangent at A' is

$$\frac{a^2 yy' \pm b^2 xx' \mp a^2 b^2}{\sqrt{a^4 y'^2 + b^4 x'^2}} = 0.$$

Hence the distance from the origin to the tangent is

$$\frac{a^2 b^2}{\sqrt{a^4 y'^2 + b^4 x'^2}} = \frac{ab}{\sqrt{\dfrac{a^2 y'^2}{b^2} + \dfrac{b^2 x'^2}{a^2}}} = \frac{ab}{b'} \quad \text{(Arts. 115, 116),}$$

and $\qquad 4\, OB' \cdot OD = 4 b' \cdot \dfrac{ab}{b'} = 4 ab.$

119. *To find the equal conjugate diameters of the ellipse.*

Equating the values of a'^2 and b'^2 (Art. 110),

$$a'^2 = \frac{a^2 b^2}{a^2 \sin^2 \gamma + b^2 \cos^2 \gamma} = b'^2 = \frac{a^2 b^2}{a^2 \sin^2 \gamma_1 + b^2 \cos^2 \gamma_1},$$

whence $\qquad a^2 \sin^2 \gamma_1 + b^2 \cos^2 \gamma_1 = a^2 \sin^2 \gamma + b^2 \cos^2 \gamma\,;$

or, transposing,

$$a^2 (\sin^2 \gamma_1 - \sin^2 \gamma) = b^2 (\cos^2 \gamma - \cos^2 \gamma_1) = b^2 (\sin^2 \gamma_1 - \sin^2 \gamma),$$

since $\qquad \cos^2 A = 1 - \sin^2 A.$

Hence $\qquad (a^2 - b^2)(\sin^2 \gamma_1 - \sin^2 \gamma) = 0, \qquad (1)$

and therefore $\sin^2 \gamma_1 = \sin^2 \gamma$. Since, in the ellipse, if γ is acute, γ_1 is obtuse, and the sines are equal,

$$\gamma_1 = 180° - \gamma \quad \text{and} \quad \tan \gamma_1 = -\tan \gamma.$$

Substituting this in $\tan \gamma \tan \gamma_1 = -\dfrac{b^2}{a^2},$

the equation of condition for conjugate diameters to the ellipse,

$$\tan \gamma = \pm \frac{b}{a}, \quad \therefore \quad \tan \gamma_1 = -\tan \gamma = \mp \frac{b}{a}.$$

Hence, *when the diameters are equal, the angles they make with the transverse axis are supplementary and the diameters fall on the diagonals of the rectangle on the axes.*

Cor. 1. If $a = b$, (1) is satisfied independently of γ and γ_1; or, *in the circle every diameter equals its conjugate.*

Cor. 2. For the hyperbola, (1) becomes

$$(a^2 + b^2)(\sin^2\gamma_1 - \sin^2\gamma) = 0,$$

which cannot be satisfied for $\sin^2\gamma_1 = \sin^2\gamma$, since in the hyperbola both angles are acute and this condition would make them coincide. Hence *the hyperbola has no equal conjugate diameters.* From $a'^2 - b'^2 = a^2 - b^2$, however, we see that if $a = b$, then $a' = b'$; or, *every diameter in the equilateral hyperbola equals its conjugate.*

SUPPLEMENTAL CHORDS.

120. Defs. Straight lines drawn from any point of an ellipse or an hyperbola to the extremities of a diameter are called **supplemental chords**.

Thus, $S''Q$, QS' (Figs. 76, 77) are supplemental chords.

121. *If a chord of an ellipse or hyperbola is parallel to a diameter, the supplemental chord is parallel to the conjugate diameter.*

Let $A''A'$ (Figs. 76, 77) be a diameter, and $S''Q$ the parallel chord. Draw the supplemental chord QS', and let x', y', be the coordinates of S', and therefore $-x'$, $-y'$, those of S''. The equation of $S''Q$ will be $y + y' = a''(x + x')$ (Art. 31), and that of $S'Q$, $y - y' = a'(x - x')$. Combining these equations by multiplication, $y^2 - y'^2 = a'a''(x^2 - x'^2)$, in which x and y are the coordinates of Q (Art. 36). But S' and Q are on the curve; hence $a^2y'^2 \pm b^2x'^2 = \pm a^2b^2$ and $a^2y^2 \pm b^2x^2 = \pm a^2b^2$; or, by subtraction, $y^2 - y'^2 = \mp \dfrac{b^2}{a^2}(x^2 - x'^2)$. Equating these two values of $y^2 - y'^2$, we have $a'a'' = \pm \dfrac{b^2}{a^2}$. But this is the condition for conjugate diameters, viz.: $\tan\gamma \tan\gamma_1 = \mp \dfrac{b^2}{a^2}$ (Arts. 90, 93). Hence if $a = \tan\gamma$, $a' = \tan\gamma_1$, and conversely.

OBLIQUE AXES.   157

Cor. 1. *To draw a tangent at a given point of the curve*, as A', draw the diameter $A'A''$ and any parallel chord as $S''Q$. Draw the chord QS' supplemental to $S''Q$. A line parallel to QS' through A' is the required tangent.

Cor. 2. *To draw a tangent parallel to a given line*, as MN, draw any chord QS' parallel to it, and the supplemental chord QS''. Then the diameter $A''A'$, parallel to $S''Q$, determines the points of tangency A'' and A'.

PARABOLA REFERRED TO OBLIQUE AXES.

122. Equation of the parabola referred to any diameter and the tangent at its vertex.

The formulæ for transforming from rectangular to oblique axes, the new origin being at O', are (Art. 22, Eq. 3)

$$x = x_0 + x_1 \cos\gamma + y_1 \cos\gamma_1, \quad y = y_0 + x_1 \sin\gamma + y_1 \sin\gamma_1. \quad (1)$$

But $\gamma = 0$, since the new axis of X is parallel to the primitive one, hence $\cos\gamma = 1$, $\sin\gamma = 0$. Also $\tan\gamma_1 = \dfrac{p}{y_0}$, since the new axis of Y is tangent to the curve at (x_0, y_0) (Art. 104, Ex. 4); hence from

$$\tan\gamma_1 = \frac{\sin\gamma_1}{\cos\gamma_1} = \frac{\sin\gamma_1}{\sqrt{1 - \sin^2\gamma_1}}$$

we have

$$\sin\gamma_1 = \frac{p}{\sqrt{y_0^2 + p^2}},$$

and therefore

$$\cos\gamma_1 = \sqrt{1 - \sin^2\gamma_1} = \frac{y_0}{\sqrt{y_0^2 + p^2}}.$$

Substituting these values in (1), they become

$$x = x_0 + x_1 + \frac{y_1 y_0}{\sqrt{y_0^2 + p^2}}, \quad y = y_0 + \frac{y_1 p}{\sqrt{y_0^2 + p^2}}.$$

Substituting these values in the equation to be transformed, $y^2 = 2px$, and remembering that, since O' is on the curve,

158   ANALYTIC GEOMETRY.

$y_0^2 = 2px_0$, we have, after omitting the subscripts of x and y,
$$y^2 = \frac{2(y_0^2 + p^2)}{p} x, \qquad (2)$$
which is the required equation.

COR. 1. $y_0 = MO' = MN \tan MNO' = p \cot \gamma_1$, $O'N$ being the normal and $MN =$ subnormal $= p$. Hence
$$\frac{2(y_0^2 + p^2)}{p} = 2p(1 + \cot^2 \gamma_1) = 2p \csc^2 \gamma_1 = \frac{2p}{\sin^2 \gamma_1},$$
and (2) may be written
$$y^2 = \frac{2p}{\sin^2 \gamma_1} x. \qquad (3)$$

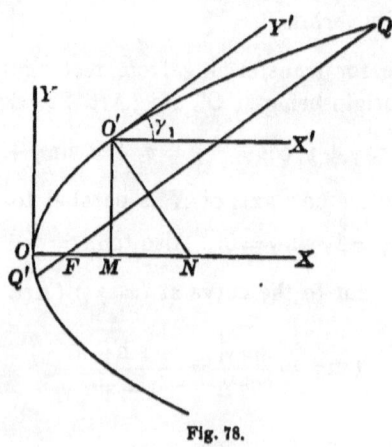

Fig. 78.

COR. 2. From the polar equation of the parabola,
$$r = \frac{p}{1 - \cos \theta};$$
making $\theta = \gamma_1$, we have
$$r = FQ = \frac{p}{1 - \cos \gamma_1};$$
making $\theta = 180° + \gamma_1$,
$$r = FQ' = \frac{p}{1 + \cos \gamma_1}.$$

OBLIQUE AXES.

Hence $$QQ' = FQ + FQ' = \frac{2p}{\sin^2 \gamma_1};$$

or, representing QQ' by $2p'$, (3) may be written

$$y^2 = 2p'x. \qquad (4)$$

Thus the equation of the parabola referred to any diameter and the tangent at its vertex is $y^2 = 2p'x$, $2p'$ being the focal chord parallel to the tangent, and becoming $2p$ when the diameter is the axis.

Cor. 3. The equation of the tangent referred to a diameter and the tangent at its vertex is $yy' = p'(x + x')$, since the only change in the process of Art. 104 is that arising from the substitution of p' for p.

123. *The squares of ordinates to any diameter of a parabola are to each other as their corresponding abscissas.*

Referred to any diameter and the tangent at its vertex the equation of the parabola is $y^2 = 2p'x$. Hence for the points P' and P'',
$$y'^2 = 2p'x', \quad y''^2 = 2p'x'';$$

or, by division, $$\frac{y'^2}{y''^2} = \frac{x'}{x''}.$$

ASYMPTOTES.

124. Equation of the hyperbola referred to its asymptotes.

The asymptotes being oblique except when the hyperbola is rectangular (Art. 92), we use the formulæ for passing to oblique axes with the same origin,

$$x = x_1 \cos \gamma + y_1 \cos \gamma_1, \quad y = x_1 \sin \gamma + y_1 \sin \gamma_1;$$

and, since the asymptotes coincide with the diagonals of the rectangle on the axes,

$$\sin \gamma = \frac{-b}{\sqrt{a^2 + b^2}}, \quad \sin \gamma_1 = \frac{b}{\sqrt{a^2 + b^2}}, \quad \cos \gamma = \cos \gamma_1 = \frac{a}{\sqrt{a^2 + b^2}}.$$

The formulæ therefore become
$$x = \frac{a}{\sqrt{a^2+b^2}}(x_1+y_1), \quad y = \frac{b}{\sqrt{a^2+b^2}}(y_1-x_1).$$

Substituting these values in $a^2y^2 - b^2x^2 = -a^2b^2$, and omitting the subscripts, we obtain
$$xy = \frac{a^2+b^2}{4}.$$

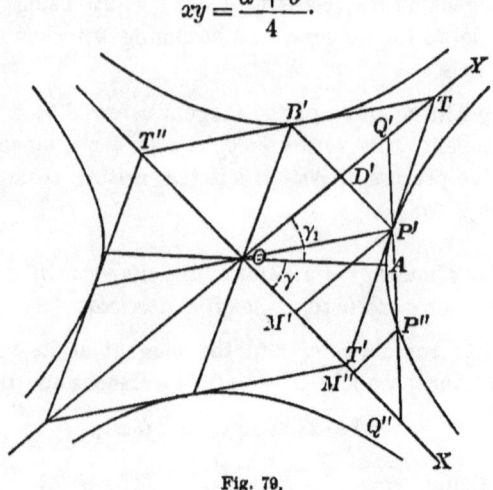

Fig. 79.

Hence the general form of the equation of the hyperbola referred to its asymptotes is $xy = m$, in which m is constant.

COR. 1. The equation of the Y-hyperbola referred to the same axes is $xy = -\frac{a^2+b^2}{4}$ (Art. 71).

COR. 2. *The equation*
$$Bxy + Dy + Ex + F = 0$$
is the general equation of the hyperbola referred to axes parallel to its asymptotes. For, transforming $xy = m$ to parallel axes by the formulæ $x = x_0 + x_1$, $y = y_0 + y_1$, we have, after dropping the subscripts, $xy + x_0y + y_0x + x_0y_0 = 0$, which is the above form. The equations of the asymptotes are evidently $x = -y_0$, $y = -x_0$.

OBLIQUE AXES.

125. *The intercepts of the secant between the hyperbola and its asymptotes are equal.*

Let P', P'' (Fig. 79), be any two points of the hyperbola,

$$y - y' = \frac{y' - y''}{x' - x''}(x - x')$$

the equation of the secant $P'P''$. Making $x = 0$ in this equation,

$$y - y' = D'Q' = \frac{y''x' - y'x'}{x' - x''}.$$

But $y'x' = y''x'' = m$, since the points are on the curve.

Hence $$D'Q' = \frac{y''x' - y''x''}{x' - x''} = y'' = P''M''.$$

Hence the triangles $P''M''Q''$, $Q'D'P'$, being equiangular, and having a side in one equal to a side in the other, are equal, and $P''Q'' = P'Q'$.

Cor. *To construct the hyperbola when the axes are given:* draw the asymptotes, the diagonals of the rectangle on the given axes, and through the extremities of the transverse axis, as A, draw $11'$, $22'$, $33'$, etc., and make $1P'$, $2P''$, $3P'''$, etc., equal respectively to $A1'$, $A2'$, $A3'$, etc. Then P', P'', P''', etc., are points of the curve. By a similar method we may construct the curve when the asymptotes and one point of the curve are given.

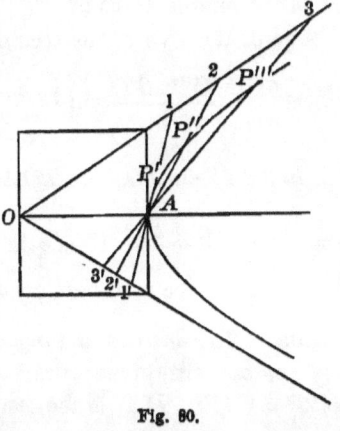

Fig. 80.

126. *The area of the triangle formed by any tangent with the asymptotes is constant, and the tangent is bisected at the point of contact.*

The equation of the secant $P'P''$ (Fig. 79) is

$$y - y' = \frac{y' - y''}{x' - x''}(x - x').$$

From the equation of the curve,

$$x'y' = x''y'' = m, \therefore y' = \frac{x''y''}{x'}.$$

The fraction $\frac{y' - y''}{x' - x''}$ therefore becomes

$$\frac{\frac{x''y''}{x'} - y''}{x' - x''} = -\frac{y''}{x'};$$

or, when P'' coincides with P', $-\frac{y'}{x'}$. Hence the equation of the tangent TT' is

$$y - y' = -\frac{y'}{x'}(x - x'),$$

and its intercepts are $y = OT = 2y'$, $x = OT' = 2x'$. Hence P' is the middle point of TT' (Art. 6).

Again, the area of the triangle OTT' is

$$\frac{OT \cdot OT'}{2} \sin TOT' = \frac{2x' \cdot 2y'}{2} \sin 2\, TOA$$

$$= 2x'y'\, 2\sin TOA \cos TOA = 4x'y'\frac{b}{\sqrt{a^2 + b^2}}\frac{a}{\sqrt{a^2 + b^2}} = ab,$$

since $x'y' = \frac{a^2 + b^2}{4}$. Hence the area of the triangle is constant and equal to the rectangle on the semi-axes.

Cor. *To construct a tangent at any point*, as P', when the asymptotes are given, draw the ordinate $P'M'$ and make $M'T' = OM'$. $P'T'$ is the tangent.

127. *Tangents at the extremities of conjugate diameters meet on the asymptotes.*

The equation of the straight line $P'B'$ (Fig. 79), the coordinates of P' being x', y', and those of B' being $\dfrac{ay'}{b}$, $\dfrac{bx'}{a}$ (Art. 116), is

$$y - y' = \frac{y' - \dfrac{bx'}{a}}{x' - \dfrac{ay'}{b}}(x - x'),$$

or
$$y - y' = -\frac{b}{a}(x - x').$$

But the equation of OT'' is $y = -\dfrac{b}{a}x$; hence $P'B'$ is parallel to the asymptote OT'. Again, the middle point of $P'B'$ is

$$\left[\tfrac{1}{2}\left(x' + \frac{ay'}{b}\right),\ \tfrac{1}{2}\left(y' + \frac{bx'}{a}\right)\right], \quad \text{or} \quad \left(\frac{bx' + ay'}{2b},\ \frac{bx' + ay'}{2a}\right),$$

which satisfy $y = \dfrac{b}{a}x$. Hence *the straight line joining the extremities of conjugate diameters is parallel to one asymptote and bisected by the other.* But the diagonals of a parallelogram bisect each other, and $P'B'$ is one diagonal of a parallelogram of which OP' and OB' are adjacent sides; hence the other diagonal coincides with the asymptote, or the tangents at P' and B' meet on the asymptote.

CHAPTER IV.

LOCI.

128. Classification of loci.

When the relation between x and y can be expressed by the six ordinary operations of algebra, viz., addition, subtraction, multiplication, division, involution, and evolution, the powers and roots in the latter cases being denoted by constant exponents, the function is called an **algebraic** function; and loci whose equations contain only algebraic functions are called **algebraic loci**.

Algebraic loci are classified according to the degree of their equations as loci of the first, second, etc., orders. We have seen that there is but one locus of the first order; that is, whose equation is of the first degree, namely, the straight line; and that all loci of the second order are conics. All loci whose equations are above the second degree are called **higher plane curves**.

A function which involves a logarithm, as $x = \log y$, is called a **logarithmic** function; one in which the variable enters as an exponent, as $y = a^x$, an **exponential** function. If a is the base of the logarithmic system, the latter function is evidently another way of expressing the former. Functions involving the trigonometrical elements, as $y = \sin x$, $x = \sin^{-1} y$, etc., are called **circular** functions. $y = \sin x$ and $x = \sin^{-1} y$ are different forms of the same relation, the former being called the **direct**, and the latter the **inverse** circular function. It may be shown that logarithmic, exponential and circular functions cannot be expressed by a finite number of algebraic functions, and for this reason they are called **transcendental** functions. A transcendental equation is one involving transcendental functions, and the locus of such an equation is called a **transcendental curve**.

The exercises which follow will afford the student practice in the production of the equation of a locus from its definition. In all cases the object is to find a relation between the given constants, x, and y; the latter being the coordinates of any point of the required locus. Any such relation, when stated in the form of an equation, will be the equation sought, whatever the axes; but the simplicity of both the solution and the resulting equation will depend upon the choice of the axes. The student will observe two cases: first, when the given conditions furnish directly a relation between x and y; second, when the conditions involve other variables; and in this case these conditions must afford a sufficient number of independent equations to permit the elimination of all the variables except x and y. Thus, if n variables are involved exclusive of x and y, the conditions must furnish $n+1$ equations.

SECTION XI.—LOCI OF THE FIRST AND SECOND ORDER.

129. 1. *Given the base of a triangle and the difference of the squares of its sides, to find the locus of the vertex.*

Let b be the given base and d^2 the constant difference. Take the base for the axis of X, and its left-hand extremity for the origin, x and y being the coordinates of the vertex. Then, by condition, $OP^2 - BP^2 = d^2$, or

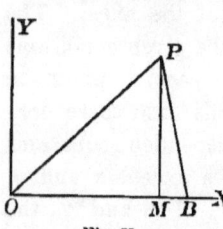

Fig. 81.

$$x^2 + y^2 - [(b-x)^2 + y^2] = d^2,$$

whence $\quad x = \dfrac{d^2 + b^2}{2b}.$

Hence the locus is a straight line parallel to Y, at a distance from it equal to $\dfrac{d^2 + b^2}{2b}$. If the triangle is isosceles, $d = 0$, and $x = \dfrac{b}{2}$. In this case the conditions furnish directly the relation between x and y.

2. *To find the locus of the middle point of a rectangle inscribed in a given triangle.*

Let $a =$ altitude of the triangle, b and c the segments of the base, the axes being taken as in the figure. Then the equations of AB and AC are known; namely,

$$\frac{x}{b} + \frac{y}{a} = 1, \text{ and } \frac{x}{c} + \frac{y}{a} = 1.$$

Now the abscissa of P is the half sum of the abscissas of Q and R; and if $y = k$ be the altitude of the rectangle, and this value be substituted for y in the above equations, we find

$$x_Q = \frac{a-k}{a} b, \quad x_R = \frac{a-k}{a} c.$$

Hence $x = $ abscissa of

$$P = \frac{x_Q + x_R}{2} = \tfrac{1}{2} \frac{a-k}{a}(b+c).$$

But the ordinate of $P = y = \frac{k}{2}$.

This condition enables us to eliminate the variable k from the above value of x; substituting therefore $k = 2y$, we have

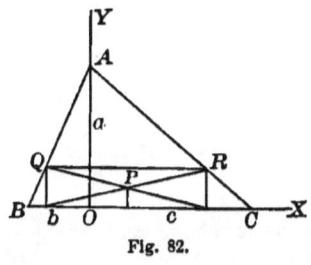

Fig. 82.

$$2ax = (a - 2y)(b+c),$$

a straight line bisecting the base and altitude, since its intercepts are $\tfrac{1}{2}(b+c)$ and $\tfrac{1}{2} a$.

3. *To find the locus of a point so moving that the square of its distance from a fixed point is in a constant ratio to its distance from a fixed line.*

Let B be the fixed point, AX the fixed line and axis of X, the axis of Y passing through B, and $OB = a$. Then, m being the constant ratio, $\frac{BP^2}{PM} = m$. But $BP^2 = x^2 + (y-a)^2$, and $PM = y$. Hence

$$y^2 + x^2 - (2a + m)y + a^2 = 0,$$

which is the equation of a circle whose centre is at $\left(0, \frac{2a+m}{2}\right)$ and whose radius is $\tfrac{1}{2}\sqrt{4am + m^2}$ (Art. 50). If the point is on the line, $a = 0$, the centre is $\left(0, \frac{m}{2}\right)$ and the radius $= \frac{m}{2}$.

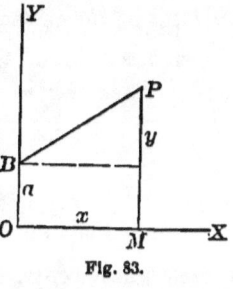

Fig. 83.

4. *The squares of the distances of a point from two fixed points are as m to n. Find the locus of the point.*

Let $(a, 0)$, $(0, b)$, be the coordinates of the fixed points, the axes being assumed to pass through them, and P any point of the locus. Then, A and B being the fixed points,

$$\frac{PB^2}{PA^2} = \frac{x^2 + (y-b)^2}{y^2 + (x-a)^2} = \frac{m}{n};$$

or, clearing of fractions and reducing,

$$y^2 + x^2 - \frac{2nb}{n-m}y + \frac{2am}{n-m}x + \frac{nb^2 - ma^2}{n-m} = 0,$$

a circle whose centre is $\left(-\dfrac{am}{n-m},\ \dfrac{nb}{n-m}\right)$, and radius is

$$\frac{1}{n-m}\sqrt{mn(a^2+b^2)}\ \text{(Art. 50)}.$$

If $b = 0$, or $a = 0$, that is, if both the points are on the same axis, the centre is on that axis. If $a = b = 0$, the centre is at the origin and $R = 0$, or the locus is a point; unless *also* $m = n$, when $R = \dfrac{0}{0}$.

5. *Find the locus of the vertex of a triangle having given the base and the sum of the squares of the sides.*

Ans. A circle whose centre is the middle point of the base.

6. *Given the base of a triangle and the ratio of its sides, find the locus of the vertex.*

Let $b =$ base of the triangle (Fig. 81), and $\dfrac{OP}{PB} = m$, the ratio. Then $OP^2 = m^2 PB^2$, or

$$x^2 + y^2 = m^2(y^2 + (b-x)^2),$$

whence
$$y^2 + x^2 + \frac{2m^2 b}{1-m^2}x - \frac{m^2 b^2}{1-m^2} = 0;$$

a circle whose centre is $\left(-\dfrac{m^2 b}{1-m^2},\ 0\right)$, and radius is $\dfrac{mb}{1-m^2}$.

7. *From one extremity A' of a diameter AA' to a circle a secant is drawn meeting the circle at P'. At P' a tangent to the circle is drawn, and from A a perpendicular to this tangent. The perpendicular produced meets the secant at P. Find the locus of P.*

Let the diameter be the axis of X and the centre the origin. Let (x', y') be the coordinates of P'; then (Art. 32)

$$y = \frac{y'}{R + x'}(R + x) \qquad (1)$$

is the equation of the secant; the equation of the tangent at P' is $yy' + xx' = R^2$, hence the perpendicular on the tangent from A is

$$y = \frac{y'}{x'}(x - R). \qquad (2)$$

Combining (1) and (2), to find P, we have

$$x = 2x' + R, \quad y = 2y'. \qquad (3)$$

But (x', y') is on the circle, hence $x'^2 + y'^2 = R^2$. Substituting in this equation the values of x' and y' from (3), we have

$$(x - R)^2 + y^2 - 4R^2 = 0,$$

a circle whose centre is at $(R, 0)$, that is, at A, and whose radius is $2R = AA'$.

8. *A line is drawn parallel to the base of a triangle, and the points where it meets the sides are joined transversely to the extremities of the base; find the locus of their intersection.* Take the sides as axes.

Ans. A straight line through the middle point of the base and the opposite vertex.

9. *Given the base and sum of the sides of a triangle, if the perpendicular be produced beyond the vertex until its whole length is equal to one of the sides, to find the locus of the extremity of the perpendicular.* *Ans. A straight line.*

10. *Given any parallelogram, and PP', QQ', lines parallel to adjacent sides. Prove that the locus of the intersection of PQ and $P'Q'$ is a diagonal of the parallelogram* (Fig. 84).

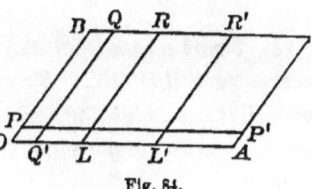

Fig. 84.

11. In Fig. 84, find the locus of the intersection of BL and PA, A and B being fixed points, and P and L subject to the condition that $OL + OB = OP + OA$.

12. *A line cuts two fixed intersecting lines so that the area of the intercepted triangle is constant. Find the locus of the middle point of the line* (see Art. 126).

Let OX, OY, be the fixed lines and axes, AOB the intercepted triangle, m the constant area, ϕ the constant angle BOA, and P the middle point of AB. Then $OM = x$, $MP = y$, and, since P is the middle point of AB, $OA = 2y$ and $OB = 2x$. Hence

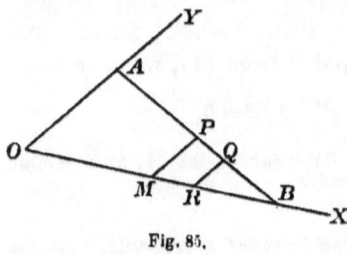

Fig. 85.

area $BOA = m = \dfrac{2x \cdot 2y \cdot \sin \phi}{2}$,

or $xy = \dfrac{m}{2 \sin \phi}$,

an hyperbola whose asymptotes are the fixed lines (Art. 124).

13. *Given two intersecting fixed lines and a fixed point. A line is drawn through the fixed point. Find the locus of the middle point of the segment intercepted by the given lines.*

Let OX, OY (Fig. 85), be the fixed lines and axes, Q the fixed point, its coordinates $OR = m$, $RQ = n$, AB the line, and P its middle point. Then, from similar triangles,

$$OA (= 2y) : OB (= 2x) :: RQ (= n) : RB (= 2x - m),$$

or $2xy - my - nx = 0$; an hyperbola passing through Q, whose asymptotes are $x = \dfrac{m}{2}$, $y = \dfrac{n}{2}$ (Art. 124, Cor. 2).

14. *From a fixed point A (Fig. 85), a line AB is drawn to meet a fixed line OX. From the intersection B, a constant distance $BR = b$ is laid off, and from R a line RQ is drawn, making a constant angle with OX, to meet AB in Q. Find the locus of Q.*

Since the angle BRQ is constant, take a parallel to QR through A for the axis of Y, and the fixed line OX for the axis of X, and let $OA = a$. Then, $OA : RQ :: OB : RB$, or $a : y :: x + b : b$; whence $xy + by - ab = 0$, an hyperbola through A, one of whose asymptotes is the fixed line and the other $x = -b$ (Art. 124, Cor. 2).

15. *To find the locus of the intersection of a perpendicular from the focus of a parabola on the normal.*

The equation of the normal is
$$y - y' = -\frac{y'}{p}(x - x'),$$
and that of the perpendicular is
$$y = \frac{p}{y'}\left(x - \frac{p}{2}\right).$$

Combining these to find the point of intersection, we find it to be
$$x = \frac{p^3 + 2x'y'^2 + 2py'^2}{2y'^2 + 2p^2}, \quad y = \frac{2px'y' + p^2y'}{2y'^2 + 2p^2}.$$

In this problem the conditions introduce the auxiliary variables x', y', the coordinates of the point of contact from which the normal is drawn. But this point is on the parabola; hence we have the additional condition $y'^2 = 2px'$. Eliminating y' by means of this equation, we have
$$x = x' + \frac{p}{2}, \quad y = \frac{\sqrt{2px'}}{2}.$$

Finally, combining and eliminating x', we have
$$y^2 = \frac{p}{2}x - \frac{p^2}{4},$$
a parabola on the same axis, whose vertex is at $\left(\frac{p}{2}, 0\right)$, and whose parameter $= \frac{1}{4}$ that of the given parabola.

16. *The locus of the intersection of the perpendicular from the focus of a parabola upon the tangent is the tangent at the vertex.*

The equation of the tangent is
$$yy' = p(x + x'),$$
and of the perpendicular upon the tangent through the focus,
$$y = -\frac{y'}{p}\left(x - \frac{p}{2}\right).$$

Combining these to find the intersection, we obtain, on eliminating y, $x(p + 2x') = 0$, which, since x' cannot be negative, is satisfied only for $x = 0$; that is, the intersection is always on Y, which is the tangent at the vertex.

How does this property enable us to find the focus when the curve and axis are given?

17. *Through any fixed point chords are drawn to a parabola. Find the locus of the intersections of the tangents to the parabola at the extremities of each chord.*

Let x_1, y_1, be the coordinates of the point through which the chords are drawn, and suppose the tangents at the extremities of one of these chords to meet at (h, k). Then the equation of the chord is (Art. 106)
$$yk = p(x + h).$$

But the chord passes through the fixed point (x_1, y_1), hence $y_1 k = p(x_1 + h)$. Now this is the equation of a straight line, in which h and k are the variables; therefore the locus of (h, k) is a straight line.

If the fixed point is the focus, $x_1 = \frac{p}{2}$, $y_1 = 0$, and the equation becomes $h = -\frac{p}{2}$. Hence *the locus of the intersection of pairs of tangents drawn at the extremities of focal chords is the directrix.*

18. *Through any fixed point chords are drawn to the ellipse (or hyperbola); to find the locus of the intersection of the tangents at the extremities of each chord.*

Let x_1, y_1, be the coordinates of the point through which the chords are drawn, the tangents at the extremities of one of

LOCI OF THE FIRST AND SECOND ORDER.

them meeting at (h, k). Then the equation of this chord is (Art. 106)
$$a^2yk \pm b^2xh = \pm a^2b^2.$$

But this chord passes through the fixed point (x_1, y_1), hence $a^2y_1k \pm b^2x_1h = \pm a^2b^2$. Now this is the equation of a straight line in which h and k are the variables; therefore the locus is a straight line.

If the fixed point is the focus, $x_1 = ae$, $y_1 = 0$, and the equation becomes $h = \dfrac{a}{e}$. Hence *the tangents at the extremities of focal chords to the ellipse and hyperbola intersect on the directrix.*

19. *The locus of the intersection of the perpendicular from the focus of an ellipse upon the tangent is the circle described on the transverse axis.*

In this problem the equation of the tangent in terms of the slope (Art. 105, Ex. 12),
$$y = mx + \sqrt{a^2m^2 + b^2},$$
is most convenient. The perpendicular upon it from the focus is
$$y = -\frac{1}{m}(x - ae).$$

From the former, $y - mx = \sqrt{a^2m^2 + b^2}$, and from the latter, $my + x = ae$. Squaring and adding,
$$(y^2 + x^2)(1 + m^2) = b^2 + m^2a^2 + a^2e^2 = b^2 + ma^2 + a^2\frac{a^2 - b^2}{a^2} = a^2(1 + m^2),$$
which eliminates m, giving $y^2 + x^2 = a^2$.

This property is also true of the hyperbola.

20. *Find the locus of the intersection of pairs of tangents to a parabola which intercept a constant length on the tangent at the vertex.*

The equations of the tangents are
$$yy' = p(x + x'), \tag{1}$$
$$yy'' = p(x + x''). \tag{2}$$

The equation of the locus will be found by combining these and eliminating x', y', x'', and y''. To effect this elimination we have the equations of condition,

$$y'^2 = 2px', \qquad (3)$$

$$y''^2 = 2px'', \qquad (4)$$

$$\frac{y' - y''}{2} = a, \text{ a constant.} \qquad (5)$$

Substituting in (1) and (2) the values of x' and x'' from (3) and (4), we have

$$yy' = px + \frac{y'^2}{2}, \qquad (6)$$

$$yy'' = px + \frac{y''^2}{2}. \qquad (7)$$

Substituting from (5) $y' = 2a + y''$ in (6) and combining the result with (7), we have $y'' = y - a$; which substituted in (7) gives $y^2 = 2px + a^2$, an equal parabola with the same axis, and vertex at $\left(-\dfrac{a^2}{2p}, 0\right)$.

21. *Parallel chords, as QQ', whose centre is C, are drawn to a circle. AA' is a diameter parallel to the chords. Find the locus of the intersection of AC with the radii through the extremities of the chords.*

Ans. A parabola whose axis is the diameter and vertex midway between O, the centre, and A.

22. *Find the locus of the intersection of tangents drawn at the extremities of conjugate diameters of an ellipse.*

Ans. $2a^2y^2 + 2b^2x^2 = 4a^2b^2$.

23. *Lines RL, $R'L'$ (Fig. 84), are drawn parallel to one side of a parallelogram and equidistant from the centre. Find the locus of the intersection of a line drawn from O through the extremity R of one of the parallels with the other, or the other produced.* *Ans. An hyperbola.*

LOCI OF THE FIRST AND SECOND ORDER.

24. *A and B are fixed points. Find the locus of P when* $\dfrac{PD^2}{AD \times DB} = a$ *constant, D being the foot of the perpendicular from P on AB.* Ans. *An ellipse.*

25. *To find the locus of the centres of all circles which pass through a given point and are tangent to a given straight line.*
Ans. *A parabola.*

26. *If a variable circle touch a fixed circle and a fixed straight line, the locus of its centre is a parabola.*

27. *Given the base of a triangle and the product of the tangents of the base angles; the locus of the vertex is an ellipse.*

28. *The base and area of a triangle is constant; the locus of the vertex is a straight line.*

SECTION XII.—HIGHER PLANE LOCI.

130. The limits of this work permit a reference to a few only of the higher plane curves possessing interesting geometric properties.

1. The cardioid. *Through any point O of a circle a secant is drawn cutting the circle in Q. Required the locus of a point P on the secant when $QP = R$, the radius of the circle.*

Let C be the centre of the circle, $CO = R$ the radius, O the pole, and OX, a tangent at O, the polar axis. Then
$$OP = OQ + QP.$$
But $OP = r$,
$OQ = OD \cos QOD = OD \sin \widehat{XOQ} = 2R \sin \theta$, and $QP = r$.

Hence $\qquad r = 2R \sin \theta + R.$ \hfill (1)

DISCUSSION OF THE EQUATION. For $\theta = 0°$, $r = R = OA$. As θ increases, r increases, and when $\theta = 90°$, $r = 3R = OB$.

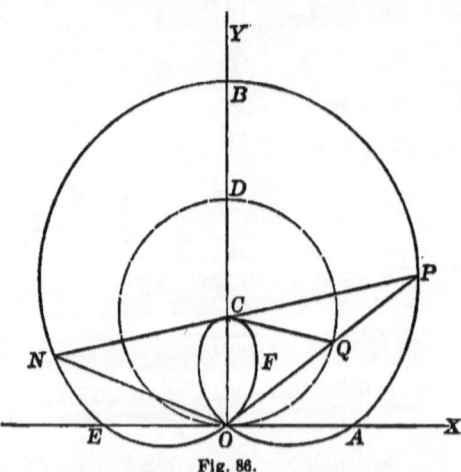

Fig. 86.

As θ increases from $90°$ to $180°$, r diminishes, and when $\theta = 180°$, $r = R = OE$. When θ passes $180°$, $\sin\theta$ becomes negative, but r remains positive until $\sin\theta = -\frac{1}{2}$, when $r = 0$; at this point $\theta = 210°$. From this value of θ, r is negative and the portion OFC is traced, r being $-R$ when $\theta = 270°$. From $\theta = 270°$ to $\theta = 360°$ the portion COA is traced, r becoming positive again when $\theta = 330°$.

RECTANGULAR EQUATION OF THE CARDIOID. Transferring to the axes YOX by the formulæ (Art. 24, Eq. 4),

$$r = \sqrt{x^2 + y^2}, \quad \sin\theta = \frac{y}{\sqrt{x^2+y^2}},$$

we have $\qquad (x^2 + y^2 - 2Ry)^2 = (x^2 + y^2) R^2, \qquad (2)$

a curve of the fourth order.

TRISECTION OF THE ANGLE. The cardioid affords a method of trisecting an angle, as follows. Let NCO be the given angle. With the vertex C as a centre describe any circle, and construct the cardioid to this circle. Only that portion of the curve in the vicinity of NC produced need be constructed. Produce NC to meet the cardioid at P, and draw PO and QC. Then the triangles CQP, CQO, are isosceles by construction. Hence
$NCO = COP + CPO = CQO + CPO = QCP + 2\,CPO = 3\,CPO$;
or $\qquad CPO = \frac{1}{3} NCO$.

2. **The conchoid.** *Through a fixed point F a line FP is*

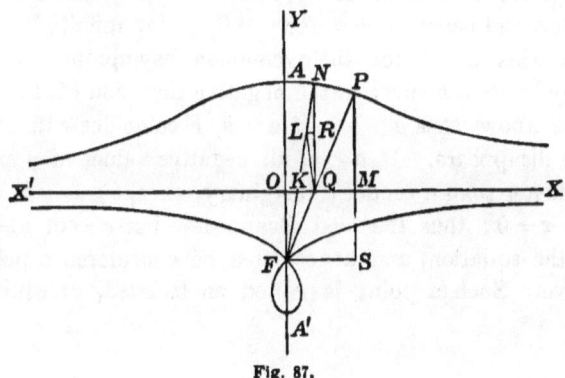

Fig. 87.

drawn cutting a fixed line XX' in Q. Required the locus of P when QP is constant.

Let $QP = a$, $FO = b$, the distance from the point to the line, and OX, OY, be the axes. Draw FS and PS parallel and perpendicular respectively to $X'X$. Then $PM : MQ :: PS : FS$;

or
$$y : \sqrt{a^2 - y^2} :: y + b : x;$$

whence
$$x^2 y^2 = (y + b)^2 (a^2 - y^2), \qquad (1)$$

a curve of the fourth order.

DISCUSSION OF THE EQUATION. Solving the equation for x, we have

$$x = \pm \frac{y+b}{y} \sqrt{a^2 - y^2}.$$

The curve is evidently symmetrical with respect to Y. When y is positive and equal to a, $x = 0$, locating the point A, which is a limit in the positive direction of Y, since x is imaginary if $y > a$. As y diminishes, x increases numerically, becoming $\pm \infty$ when $y = 0$; hence the curve has infinite branches in the first and second angles with $X'X$ for their common asymptote. Since x is real for negative values of y less than a numerically, there is a branch in the third and fourth angles. When $y = -a$, or $-b$, $x = 0$, locating A' and F; and as x has two values numerically equal with opposite signs for values of y between $-a$ and $-b$, the locus between these values is an oval. When y is negative and numerically less than b, x increases as y diminishes and becomes $\pm \infty$ for $y = 0$, giving infinite branches with the axis of X for their common asymptote. $y = -a$ evidently limits the curve in the negative direction of Y.

In the above case $a > b$. If $a = b$, F coincides with A' and the oval disappears. If $a < b$, all negative values of y numerically greater than a render x imaginary, except $y = -b$, which renders $x = 0$; thus the oval disappears, but $x = 0$, $y = -b$, satisfy the equation, and hence must be considered a point of the curve. Such a point is called an **isolated**, or **conjugate** **point**.

POLAR EQUATION OF THE CURVE. The polar equation may be obtained by transformation, or directly from the figure, thus: Let F be the pole and FS the polar axis; then

$$r = FP = FQ + QP = FO \sec\theta + a,$$
or $$r = b \operatorname{cosec}\theta + a. \qquad (2)$$

TRISECTION OF THE ANGLE. The conchoid also affords a method of trisecting the angle, as follows: Let AFP be the given angle. Draw any line OX perpendicular to one side, and with F as a fixed point, OX a fixed line, and $PQ = 2FQ$, construct the arc of a conchoid. Only that portion of the curve included within the given angle need be drawn. From Q draw QN perpendicular to OX and join its intersection with the conchoid, N, with F. Bisect QN at R, draw RL parallel to OX, and join L with Q. Then the triangles LNQ, LFQ, are isosceles; for, since $QR = RN$,

$$KL = LN = LQ = \tfrac{1}{2} QP = FQ.$$

Hence the angle

$$AFN = FNQ = LQN = \tfrac{1}{2} FLQ = \tfrac{1}{2} LFQ, \text{ or } AFN = \tfrac{1}{3} AFP.$$

MECHANICAL CONSTRUCTION. The conchoid may be constructed mechanically as follows: Let AA', XX', be two fixed rulers, the latter having a groove on its upper surface. Let FP be a third ruler, having a peg Q fixed on its under side, which is also grooved to slide on a peg at F. A pencil at P will trace the curve.

3. **The cissoid.** *Pairs of equal ordinates are drawn to the diameter of a circle, and through one extremity of the diameter a line is drawn through the intersection of one of the ordinates with the circle. Find the locus of the intersection of this line with the equal ordinate or that ordinate produced.*

Let OD be the diameter to which the ordinates are drawn, and the axis of X, the tangent to the circle at O being the axis of Y. Let QM, $Q'M'$, be equal ordinates. Through O

draw OQ (or OQ'); then P' (or P) is a point of the locus. From similar triangles,

$$OM:MP::OM':M'Q';$$

or, R being the radius of the circle,

$$x:y::2R-x:\sqrt{x(2R-x)},$$

whence $y^2(2R-x)=x^3.$ (1)

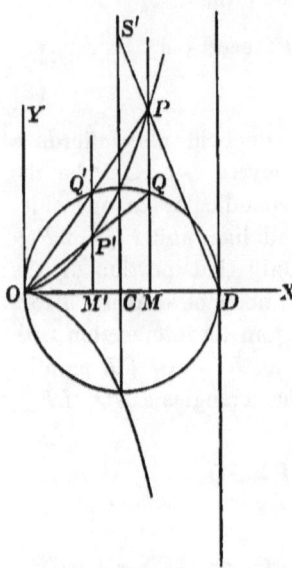

Fig. 88.

DISCUSSION OF THE EQUATION. Solving the equation for y,

$$y=\pm\sqrt{\frac{x^3}{2R-x}},$$

which shows that the curve is symmetrical with respect to X. If x is negative or greater than $2R$, y is imaginary; hence the limits along X are zero and $2R$. As x increases, y increases, and when $x=2R$, $y=\pm\infty$; hence the curve has two infinite branches which have the tangent at D for a common asymptote.

POLAR EQUATION OF THE CURVE. Let O be the pole and OX the polar axis. Substituting in (1) the values

$$x=r\cos\theta,\ y=r\sin\theta\ \text{(Art. 23, Eq. 4)},$$

we have $\quad r^2\sin^2\theta(2R-r\cos\theta)=r^3\cos^3\theta.$

But $\quad \sin^2\theta=1-\cos^2\theta;$

hence $\quad 2R-2R\cos^2\theta-r\cos\theta=0.$

Substituting $\dfrac{1}{\sec\theta}$ for $\cos\theta$, and remembering that

$$\sec^2\theta-1=\tan^2\theta,$$

we obtain finally $\quad r=2R\sin\theta\tan\theta.$ (2)

HIGHER PLANE LOCI.

DUPLICATION OF THE CUBE. This curve affords a method for finding the edge of a cube whose volume shall be n times that of a given cube, as follows: C being the centre of the circle, take $CS = nCD$, and draw SD intersecting the cissoid at P. Then the ordinate $PM = n \cdot MD$, and the cube whose edge is PM is n times the cube whose edge is OM. For, P being a point of the cissoid, we have from its equation,

$$PM^2 = \frac{OM^3}{MD} = \frac{OM^3}{\frac{1}{n}PM}, \quad \text{or} \quad PM^3 = n \cdot OM^3.$$

Let c be the edge of the given cube. Find c', so that

$$c' : c :: PM : OM, \quad \text{or} \quad c'^3 : c^3 :: PM^3 : OM^3.$$

Then $c'^3 = nc^3$, for $PM^3 = n \cdot OM^3$. To duplicate the cube, make $n = 2$; that is, take $CS = 2\,CD$.

4. The lemniscate. *To find the locus of the intersection of the perpendicular from the centre of an hyperbola upon the tangent.*

Assuming the form $\quad y = mx + \sqrt{a^2 m^2 - b^2}$

of the equation of the tangent (Art. 105, Ex. 12), that of the perpendicular is
$$y = -\frac{1}{m}x.$$

Substituting the value of m from the latter in the former, we have
$$(y^2 + x^2)^2 = a^2 x^2 - b^2 y^2, \tag{1}$$
a curve of the fourth order.

POLAR EQUATION OF THE CURVE. The formulæ for transformation being

$$x^2 + y^2 = r^2, \quad x = r\cos\theta, \quad y = r\sin\theta,$$

we have

$$r^2 = a^2 \cos^2\theta - b^2 \sin^2\theta,$$

or $\quad r^2 = a^2 - (a^2 + b^2)\sin^2\theta. \quad (2)$

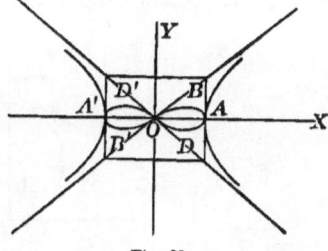

Fig. 89.

DISCUSSION OF THE POLAR EQUATION. If $\theta = 0$, $r = \pm a$, locating A and A'. As θ increases, r diminishes numerically till $\sin\theta = \dfrac{a}{\sqrt{a^2+b^2}}$; that is, till r coincides with the asymptote, when $r = \pm 0$, and the portions ABO, $A'B'O$, are traced. r then becomes imaginary, and remains so till $\sin\theta = \dfrac{a}{\sqrt{a^2+b^2}}$ again, θ being in the second quadrant and r coinciding with the other asymptote. From $\theta = \sin^{-1}\dfrac{a}{\sqrt{a^2+b^2}}$ to $\theta = 180°$, $\sin^2\theta$ is diminishing and r increasing numerically, the portions $OD'A'$, ODA, being traced, r being $\pm a$ when $\theta = 180°$.

If the hyperbola is equilateral, $a = b$, and (1) becomes

$$(y^2 + x^2)^2 = a^2(x^2 - y^2),$$

and (2) in like manner becomes

$$r^2 = a^2(\cos^2\theta - \sin^2\theta) = a^2 \cos 2\theta.$$

In this case r is imaginary for all values of θ between 45° and 135°.

5. The witch. *The ordinate to the diameter of a circle is produced till its entire length is to the diameter as the ordinate is to one of the segments with which it divides the diameter, these segments being taken on the same side. Find the locus of the extremity of the produced ordinate.*

Let OAB be the circle, R its radius, OD the diameter and axis of Y, O the origin, and the tangent at O the axis of X.

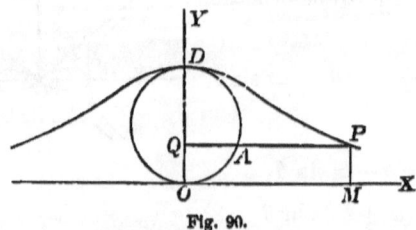

Fig. 90.

HIGHER PLANE LOCI. 183

Then, if P is a point of the curve,

$$PQ:DO::AQ:QO,$$

or $\qquad x:2R::\sqrt{2Ry-y^2}:y,$

whence $\qquad x^2y^2 = 4R^2(2Ry-y^2),\qquad (1)$

a locus of the fourth order. Let the student discuss the equation.

6. *To find the locus of the intersection of the perpendicular from the vertex of a parabola upon the tangent.*

$y^2 = 2px$ being the equation of the parabola, that of the required locus is

$$y^2 = \frac{-x^3}{\frac{p}{2}+x},$$

a cissoid, the diameter of whose circle $=\dfrac{p}{2}$.

7. *Given two fixed points F and F', to find the locus of P such that $PF \times PF' = \left(\dfrac{FF'}{2}\right)^2$.*

Let FF' be the axis of X, and the origin in the middle point of FF'. Then $(y^2+x^2)^2 = 2c^2(x^2-y^2)$ is the required locus, in which $c = \tfrac{1}{2}FF'$. The locus is the lemniscate (see Ex. 4), the hyperbola being rectangular, and $c = \dfrac{a}{\sqrt{2}}$.

8. *The corner of a rectangular sheet of paper is folded over so that the sum of the folded edges is constant. Find the locus of the corner.*

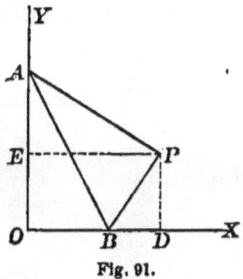

Fig. 91.

By condition, $OB = BP$, $OA = AP$, the angle at P is a right angle, and $AP + PB = a$, a constant. But

$$AP^2 = AO^2 = AE^2 + EP^2 = (AO-y)^2 + x^2,$$

$$\therefore x^2 + y^2 = 2AO \cdot y = 2AP \cdot y.$$

184 ANALYTIC GEOMETRY.

Also $PB^2 = OB^2 = PD^2 + BD^2 = y^2 + (x - OB)^2$,

$\therefore x^2 + y^2 = 2\,OB \cdot x = 2\,PB \cdot x.$

Substituting the values of AP and PB from these equations in $AP + PB = a$, we have

$$(x^2 + y^2)(x + y) = 2\,axy,$$

a locus of the third order.

SECTION XIII.—TRANSCENDENTAL CURVES.

131. 1. **The logarithmic curve.**

The equation of this curve is $x = \log y$. Assuming the form $y = a^x$, in which a is the base of the logarithmic system, we observe that when $x = 0$, $y = 1$, whatever the base; hence all

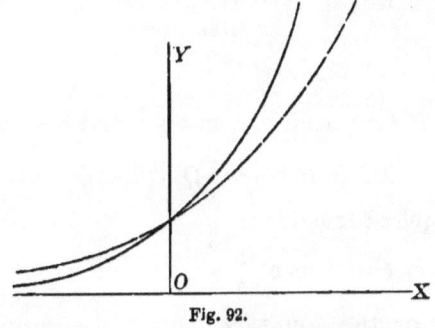

Fig. 92.

logarithmic curves cut the axis of Y at a distance unity from the origin. Again, since negative numbers have no logarithms, y cannot be negative, hence these curves lie wholly above X. If x is positive and increasing, y increases, but more rapidly than x, and the more rapidly as the base is greater; hence the curve departs rapidly from X in the first angle, and the more so as the base is greater. If x is negative, then $y = a^{-x} = \dfrac{1}{a^x}$, from which we see that as x increases numerically, y decreases, and the more rapidly as the base is greater, but becomes zero only when $x = -\infty$; hence the curve approaches X in the second angle, and that axis is an asymptote.

2. **The cycloid.** *To find the locus of a point in the circumference of a circle which rolls without sliding along a fixed straight line.*

186   ANALYTIC GEOMETRY.

Let OX be the fixed straight line and axis of X, O the initial position of the generating point and origin, r the radius of the circle, and P any point of the locus. Then $OM = ON - MN$.

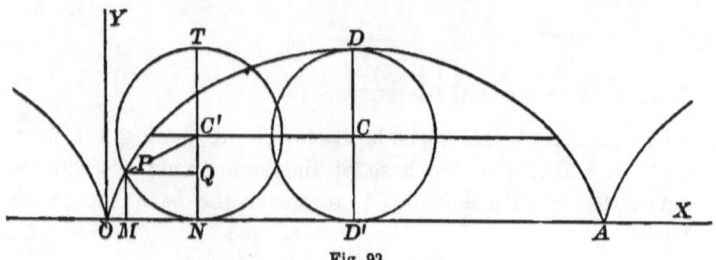

Fig. 93.

But $\qquad OM = x$,

$ON = $ arc $PN = $ versin$^{-1} QN$ to the radius $r = r$ versin$^{-1} \dfrac{y}{r}$,

and $\qquad MN = PQ = \sqrt{NQ \cdot QT} = \sqrt{2ry - y^2}$.

Hence the required equation is

$$x = r \text{ versin}^{-1} \frac{y}{r} - \sqrt{2ry - y^2}. \qquad (1)$$

DISCUSSION OF THE EQUATION. Since x is imaginary if y is negative, the curve lies wholly above X. If

$$y = 0, \quad x = r \text{ vers}^{-1} 0 = 0, \ \pm 2\pi r, \ \pm 4\pi r, \text{ etc.,}$$

or there are an infinite number of arcs equal to ODA on each side of Y, OA being equal to the circumference of the circle.

If $\qquad y = 2r, \quad x = r \text{ vers}^{-1} 2 = \pm \pi r, \ \pm 3\pi r, \text{ etc.,}$

locating D, and the corresponding points on the other arcs.

Defs. DD' is called the **axis** of the cycloid, OA the **base**, and O, A, etc., the **vertices**. To put the generating circle in position for any point, as P, draw CC' parallel to the base through the centre of the axis, and with P as a centre and a radius $= CD$ describe an arc cutting the parallel in C'. Then C' is the required centre. If the angle

$$PC'N = \phi, \quad ON = \text{arc } PN = r\phi,$$

and
$$\left.\begin{array}{l} x = r(\phi - \sin\phi) \\ y = r(1 - \cos\phi) \end{array}\right\}, \qquad (2)$$

which are called the equations of the cycloid, and are more useful than Eq. (1) in determining its properties.

If any other point than P of the radius of the rolling circle be the generating point, the resulting curve is called the *prolate*, or *curtate* cycloid, according as the generating point is within or without the circle. The locus of a point on the circumference of a circle rolling without sliding on the circumference of another is called an *epicycloid*, or *hypocycloid*, according as the circle rolls on the exterior or interior of the fixed circle; if the generating point is not on the circumference of the rolling circle, the curve is called an *epitrochoid* or *hypotrochoid*. The general term applied to the locus generated by a point of a rolling curve is *roulette*.

The circular functions. A series of transcendental curves is obtained by assuming the ordinate some trigonometrical function of the abscissa, as $y = \sin x$, $y = \cot x$, etc. The length of the arc corresponding to any value of x given in degrees may be found as follows: The length of the arc of 180° in the circle whose radius is unity being 3.1416, the length of any other arc, as that of 10°, will be $\frac{10}{180}(3.1416) = .1745$; this distance being laid off on the axis of X, the corresponding value of y may be taken from the table of natural sines, tangents, etc. The curve may be drawn, however, with sufficient accuracy by observing the general change in the function as the arc increases.

3. $y = \sin x$. When $x = 0°$, $y = 0$, hence the curve passes through the origin. As x increases, y increases, reaching its greatest value $y = 1$ when $x = 90°$, locating A. From $x = 90°$ to $x = 180°$, y decreases from 1 to 0, becoming negative when $x > 180°$, and reaching its least value $y = -1$ when $x = 270°$, locating C. From $x = 270°$ to $x = 360°$, y is negative and decreasing numerically, becoming zero again for $x = 360°$. It is evident that as x varies from 360° to 720°, a like portion will

be traced, as also when x is negative; hence the curve consists of an infinite number of arcs equal to $OABCD$, and extends

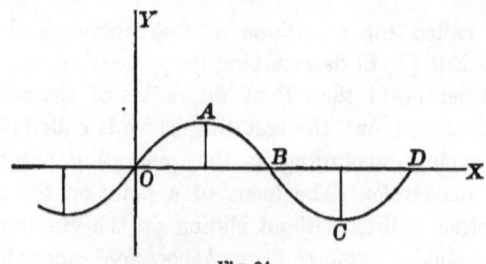

Fig. 94.

without limit along X to $\pm \infty$. The curve is sometimes called the **sinusoid**.

4. $y = \cos x$. Let the student trace the curve.

5. $y = \tan x$. When $x = 0°$, $y = 0$. As x increases, y increases, becoming ∞ when $x = 90°$. When x passes $90°$, y is negative, and remains negative till $x = 180°$, decreasing numerically from ∞ to 0. From $x = 180°$ to $x = 270°$, y is positive and increasing, becoming ∞ when $x = 270°$, etc. The curve consists of an infinite number of branches equal to AOB, on either side of the origin, having for asymptotes the lines $x = \pm \dfrac{\pi}{2}$, $x = \pm \dfrac{3}{2}\pi$, etc.

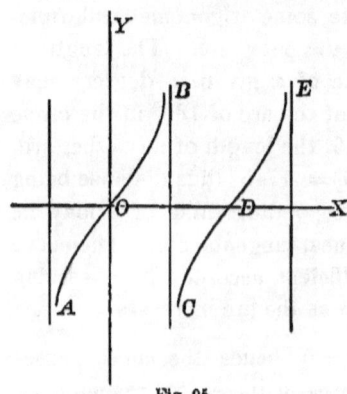

Fig. 95.

6. $y = \cot x$. Let the student trace the curve.

7. $y = \text{versin } x$. The versine being always positive, the curve lies wholly above X, its limits along Y being 0 and 2. Let the student trace the curve; also:

8. $y = \text{coversin } x$.

TRANSCENDENTAL CURVES. 189

9. $y = \sec x$. The curve is given in the figure. Let the student discuss the equation, and also trace the curve:

10. $y = \cosec x$.

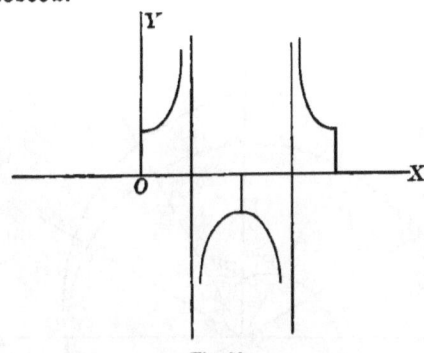

Fig. 96.

Spirals. *The locus of a point receding from a fixed point along a straight line, which revolves about the fixed point in the same plane, is called a plane spiral.*

The fixed point is called the **pole**, and that portion of the spiral traced during one revolution of the line is called a **spire**. The polar equations of many of the spirals may be derived from the general form $r = a\theta^n$, by assigning different values to n.

11. SPIRAL OF ARCHIMEDES. The equation of this spiral is obtained from the general form by making $n = 1$; whence

$$r = a\theta. \qquad (1)$$

From this equation $\dfrac{r}{\theta} = a$; since the ratio of the radius vector to the vectorial angle is constant, the spiral may be defined as traced by *a point which recedes uniformly from, while the line revolves uniformly about, the pole.* Assuming as a unit radius the value of r when the line has made one revolution, we have

$$1 = a \cdot 2\pi; \therefore a = \frac{1}{2\pi},$$

and Eq. (1) becomes $\qquad r = \dfrac{\theta}{2\pi} \qquad (2)$

when $\theta = 0$, $r = 0$, or the spiral begins at the pole. The distance between any two consecutive spires measured on the same radius is the same and equal to the unit radius, called the radius

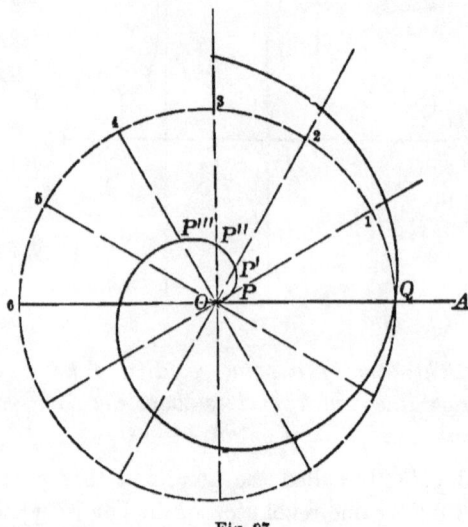

Fig. 97.

of the *measuring circle*. r increases uniformly with θ, but is ∞ only where $\theta = \infty$.

To construct this spiral, let O be the pole and OA the polar axis. Through the pole draw any number of straight lines making equal angles with each other, say 30°, or $\dfrac{\pi}{6}$. Then when $\theta = \dfrac{\pi}{6}$, $r = \tfrac{1}{12} = OP$. Having laid off $OP = \tfrac{1}{12}$ on $O1$, make $OP' = 2\,OP$, $OP'' = 3\,OP$, etc. Then $OPP'P''$, etc., is the spiral. OQ is the radius of the measuring circle.

12. The reciprocal or hyperbolic spiral. The equation of this spiral is obtained from the general form by making $n = -1$; whence

$$r = \frac{a}{\theta}. \qquad (1)$$

In this spiral the radius vector evidently varies inversely as the angle. Assuming as before that r is unity when $\theta = 2\pi$, we have $a = 2\pi$, or

$$r = \frac{2\pi}{\theta}. \tag{2}$$

When $\theta = 0$, $r = \infty$, and as θ increases, r diminishes, but is

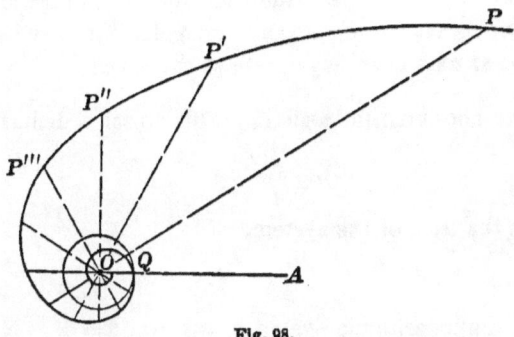

Fig. 98.

zero only when $\theta = \infty$; hence there are an infinite number of spires between the measuring circle and the pole.

To construct the spiral, draw the lines making equal angles with each other, as before. If the angle is 30°, then when $\theta = \frac{\pi}{6}$, $r = 12 = OP$. Make $OP' = \frac{OP}{2} = 6$, $OP'' = \frac{OP}{3} = 4$, etc., and draw $PP'P''\cdots$.

13. The lituus. This spiral corresponds to $n = -\frac{1}{2}$ in the general equation. Hence its equation is

$$r = \frac{a}{\sqrt{\theta}}, \tag{1}$$

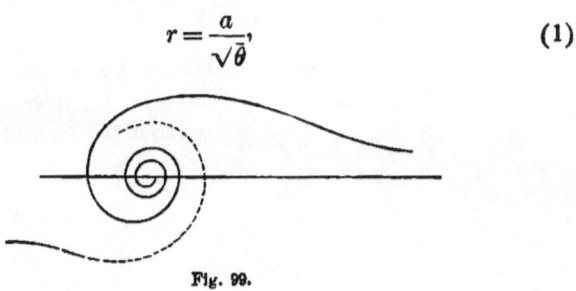

Fig. 99.

or, if $r = 1$ when $\theta = 2\pi$, $\therefore a = \sqrt{2\pi}$,
$$r^2 = \frac{2\pi}{\theta}. \tag{2}$$

For every value of θ, r has two values, one positive and one negative, as shown in the figure. If $\theta = 0$, $r = \infty$, and $r = 0$ only when $\theta = $; there are thus an infinite number of spires between the measuring circle and the pole. It may be shown that the polar axis is an asymptote to the spiral.

14. THE LOGARITHMIC SPIRAL. This spiral is defined by the equation
$$\log r = a\theta, \tag{1}$$
or, if b be the base of the system,
$$r = b^{a\theta}. \tag{2}$$

Whatever the logarithmic system $r = 1$ for $\theta = 0$. Hence, if $OA = 1$, all logarithmic spirals pass through A; and OA may be taken as the radius of the measuring circle. From (2) we see that as θ increases, r increases rapidly, and the more so as the base is greater; and diminishes rapidly if θ is negative, but is zero only when $\theta = -\infty$. Also, if $\theta = \infty$,
$$r = \infty.$$

Hence there are an infinite number of spires within and without the measuring circle.

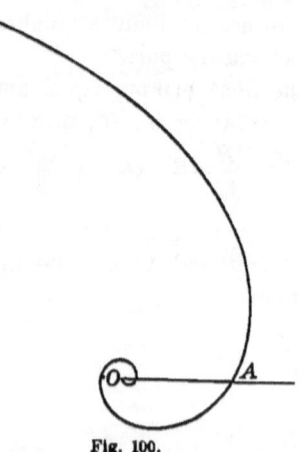

Fig. 100.

PART II.

SOLID ANALYTIC GEOMETRY.

CHAPTER V.

THE POINT, STRAIGHT LINE, AND PLANE.

SECTION XIV.—INTRODUCTORY THEOREMS.

132. Defs. 1. *By the angle between two straight lines not in the same plane is meant the angle between any two intersecting parallels.* Hence, if through any point of one of the lines, a parallel is drawn to the other, the angle between the first and the parallel is the angle between the two lines. Thus, let PQ and KL be any two straight lines which neither intersect nor are parallel. Through any point of PQ, as P, draw PH parallel to KL. Then HPQ is the angle between PQ and KL.

2. *The foot of a perpendicular from a point upon a plane is called the* **projection** *of the point on the plane.* Thus, if P be any point, AB any plane, and the perpendicular to the plane through P meets the plane at M, M is the projection of P on AB.

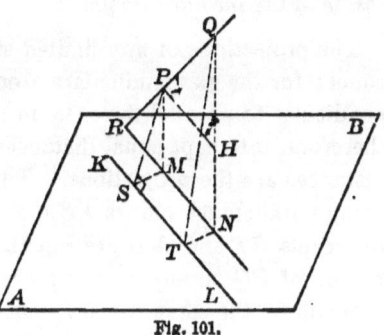

Fig. 101.

3. *The foot of a perpendicular from a point on a line is the* **projection** *of the point on the line.* Thus, if KL be any straight line and the perpendicular from P meets the line at S, S is the projection of P on KL.

4. *The projection of a straight line of limited length upon a plane is the line joining the feet of the perpendiculars from the extremities of the line upon the plane.* Thus, if PQ be any limited straight line, AB any plane, PM, QN, perpendiculars to the plane meeting it at M and N, MN is the projection of the line PQ upon the plane AB. Since the perpendiculars PM and QN determine a plane through PQ perpendicular to AB, the projection of a straight line upon a plane may also be defined as the intersection of a plane through the line, perpendicular to the given plane, with the latter.

5. *The projection of a limited straight line upon another straight line is that portion of the latter intercepted by the projections of the extremities of the former.* Thus, PS and QT being the perpendiculars from P and Q on KL, ST is the projection of PQ on KL.

133. *The projection of a limited straight line upon another straight line is equal to the length of the line multiplied by the cosine of the included angle.*

The projections of any limited straight line upon parallels are equal; for the perpendiculars from its extremities, being perpendicular to parallel lines, lie in parallel planes; these planes, therefore, intercept equal distances on the parallels, and these distances are the projections. Thus, in Fig. 101, KL and MN being parallel, the planes PSM and QTN are parallel, and the intercepts ST and MN are equal. Hence, if we find the projection of PQ on any one of a system of parallels, this projection will be the same for all. Draw PH parallel to MN. The angle between PQ and any parallel to PH is (Art. 132, 1) HPQ, and $HP = PQ \cos HPQ = MN = ST =$ etc. Hence the proposition.

Cor. Since the angle between PQ and $AB = NRQ = HPQ$, *the projection of a limited straight line upon a plane is equal to the length of the line multiplied by the cosine of the angle which the line makes with the plane.* Thus $MN = PQ \cos NRQ$.

INTRODUCTORY THEOREMS.

134. *If AD be the straight line joining A with D, and AB, BC, CD, straight lines forming a broken line from A to D, then the algebraic sum of the projections of the latter upon any straight line OX is equal to the projection of the former on the same line.*

Draw from A, B, C, and D, the perpendiculars to OX, meeting OX in A', B', C', D', respectively. It is evident that as A moves to D along the broken line $ABCD$, the foot A', of the perpendicular from A, moves along

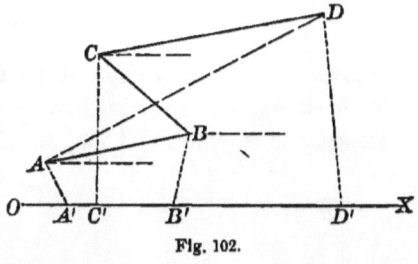

Fig. 102.

$A'D'$, to the right or the left, according as the angle between the direction of motion of A and OX is acute or obtuse. At A, B, and C, draw parallels to OX, and denote the angles made by AB, BC, CD, and AD with these parallels by α, β, γ, and δ. As the points are chosen in the figure, in passing from B to C, B' moves to the left along OX, the angle B being obtuse. Now the projection on OX of AB is (Art. 133)

$$A'B' = AB \cos \alpha;$$

that of BC is $\qquad B'C' = BC \cos \beta;$

that of CD is $\qquad C'D' = CD \cos \gamma;$

and the algebraic sum of these projections is

$$A'B' - B'C' + C'D',$$

since $\cos \beta$ is negative. But this sum is $A'D'$, which is also equal to $AD \cos \delta$, or the projection of AD on OX.

Hence $\qquad AB \cos \alpha + BC \cos \beta + CD \cos \gamma = AD \cos \delta.$

The same will evidently be true if we take any number of lines between A and D. Hence, *if two given points are joined by a broken line, the algebraic sum of the projections of its parts upon any given straight line is equal to the projection on the same line of the straight line joining the two given points.*

SECTION XV. — THE POINT.

135. Position of a point in space. The position of any point in space may be determined by referring it to three fixed planes meeting in a point. Thus, if XOY, YOZ, ZOX, be three planes meeting at O, and intersecting each other in the lines OX, OY, OZ, the position of P, relatively to these planes, will be known when its distances PQ, PR, PS, from each, measured parallel to the other two, are known. The three planes are called the **Coordinate Planes**, their three lines of intersection the **Coordinate Axes**, their common point the **Origin**, and the distances PQ, PR, PS, the **Coordinates** of P.

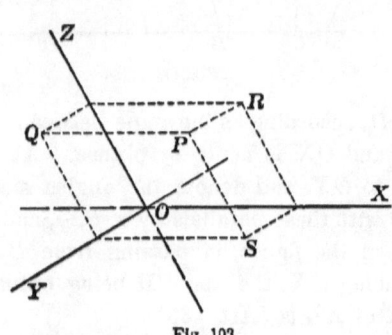

Fig. 103.

If the planes, and consequently the axes, are at right angles to each other, the coordinates are said to be **rectangular**; otherwise they are **oblique**. Use will be made only of rectangular coordinates, and they will therefore be in all cases the *perpendicular distances* of the point from the coordinate planes.

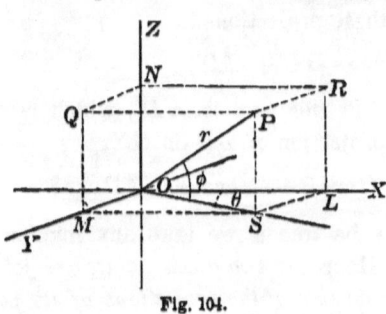

Fig. 104.

It is customary to assume the axes as in the figure, the plane XOY being horizontal and the axis OZ vertical. For brevity, the coordinate planes will be referred to as the planes XY, YZ, and ZX, and the coordinate axes as the axes of X, Y, and Z. The coordinates PQ, PR, PS, are represented by the letters x, y, z, corresponding to the axes to which they are parallel.

The coordinate planes divide space into eight right triedral angles which are numbered as follows: the *first* lies above XY, to the right of YZ, and in front of ZX; the *second* to the left of the first; the *third* behind the second; the *fourth* behind the first; the *fifth*, *sixth*, *seventh*, and *eighth*, lying under the first, second, third, and fourth, in order.

If we extend to Z the convention of signs already adopted for X and Y, the positive direction of Z being upward, it is evident that while the absolute values of x, y, z, may be the same for different points, their signs will determine in which of the eight angles any given point lies, and that a point will thus be completely determined when its coordinates are given in magnitude and sign.

136. Equation of a point. The position of a point may be designated by the equations $x = a$, $y = b$, $z = c$; or by the notation (a, b, c), the coordinates being written in the order x, y, z. To construct the point (x, y, z), construct the point (x, y), S of Fig. 104, in the plane XY, and at S erect the perpendicular $SP = z$.

EXAMPLES. 1. In what angle is the point $(a, -b, -c)$?

2. Write the coordinates of a point in each of the eight angles.

3. In what plane is the point $(a, b, 0)$?

4. Write the coordinates of a point in each of the three coordinate planes.

5. To what plane is the point (x, y, c) restricted?
    Ans. To a plane parallel to XY, at a distance c from it.

6. What are the coordinates of points in a plane parallel to YZ at a distance a from it?

7. Where is the point $(x, -b, z)$?

8. On what axis is the point $(a, 0, 0)$?

9. Write the coordinates of a point on each of the three coordinate axes.

10. What are the coordinates of the origin?

137. Distance between two given points.

Let x', y', z', be the coordinates of P'; x'', y'', z'', those of P''.

Then $\qquad P'P'' = \sqrt{P'K^2 + KP''^2}$.

But $\qquad P'K^2 = P'H^2 + HK^2$.

Hence $\qquad P'P'' = \sqrt{P'H^2 + HK^2 + KP''^2}$.

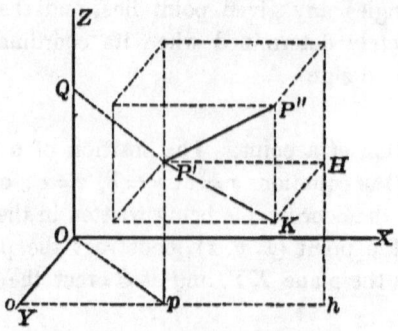

Fig. 105.

Now $\qquad P'H = ph = oh - op = x'' - x'$.

Similarly, $\qquad HK = y'' - y'$, $KP'' = z'' - z'$.

Hence, if $\qquad P'P'' = D$,
$$D = \sqrt{(x''-x')^2 + (y''-y')^2 + (z''-z')^2}.$$

COR. 1. If one of the points, as P'', is at the origin,
$$x'' = y'' = z'' = 0.$$

Hence *the distance of a point (x', y', z') from the origin is*
$$D = \sqrt{x'^2 + y'^2 + z'^2}.$$

THE POINT.

COR. 2. If $z' = 0$, that is, if P' is at p in the plane XY,
$$D = \sqrt{x'^2 + y'^2}$$
is the distance of p from the origin = distance of P' from the axis of Z. Hence the distances of (x', y', z') from the axes of X, Y, Z, are
$$\sqrt{y'^2 + z'^2}, \quad \sqrt{x'^2 + z'^2}, \quad \sqrt{x'^2 + y'^2},$$
respectively.

138. Polar coordinates of a point.

The position of a point in space may also be determined by polar coordinates. For this purpose assume any fixed plane, as XY (Fig. 104), and any fixed line in that plane, as OX, O being the pole. Then the position of P will be determined when we know OP, its distance from the pole; the angle SOP, which OP makes with the plane XY; and the angle SOX which the line SO makes with X. OP is called the radius vector of P and is represented by r, OS is the projection (Art. 134, 4) of OP on XY, the angle SOP being represented by ϕ, and ZOS by θ. The point P may then be designated as the point (r, θ, ϕ), θ and ϕ determining its direction, and r its distance, from the pole O.

139. Relations between polar and rectangular coordinates.

In Fig. 104, we have,

$$x = OL = OS \cos\theta = OP \cos\phi \cdot \cos\theta = r \cos\phi \cos\theta, \quad (1)$$
$$y = LS = OS \sin\theta = OP \cos\phi \cdot \sin\theta = r \cos\phi \sin\theta, \quad (2)$$
$$z = PS = OP \sin\phi = r \sin\phi; \quad (3)$$

the rectangular in terms of the polar coordinates.

From Art. 137, Cor. 1, we have
$$OP = r = \sqrt{x^2 + y^2 + z^2}; \quad (4)$$
from the triangle OLS,
$$\tan\theta = \frac{SL}{OL} = \frac{y}{x}; \quad (5)$$

from the triangle SOP,

$$\tan \phi = \frac{PS}{OS} = \frac{z}{\sqrt{x^2 + y^2}}; \qquad (6)$$

the polar in terms of the rectangular coordinates.

140. Direction angles and cosines.

The angles made by any straight line with the axes are called its **direction-angles.** Since parallel lines make equal angles

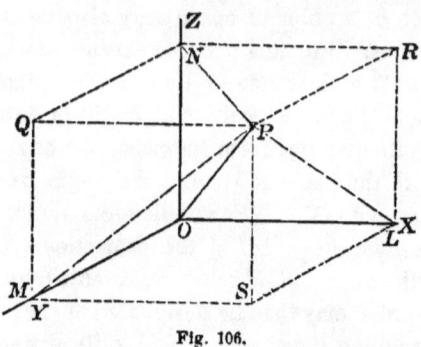

Fig. 106.

with the axes, draw OP, parallel to the given line, through the origin. Then $LOP = \alpha$, $MOP = \beta$, $NOP = \gamma$, are the direction-angles of OP, or of any parallel to OP, and are always estimated *from the positive directions of the axes*. The cosines of the direction-angles are called the **direction-cosines.**

141. *The sum of the squares of the direction-cosines of any straight line is unity.*

Join P, Fig. 106, with L. Then, since the plane PSL is perpendicular to OX, the triangle PLO is right-angled at L, and
$$OL = x = OP \cos LOP = r \cos \alpha;$$
similarly, drawing PM and PN,
$$OM = y = OP \cos MOP = r \cos \beta,$$
$$ON = z = OP \cos NOP = r \cos \gamma.$$

Squaring and adding,

$$OL^2 + OM^2 + ON^2 = r^2(\cos^2\alpha + \cos^2\beta + \cos^2\gamma).$$

But the first member is r^2 (Art. 137, Cor. 1). Hence

$$\cos^2\alpha + \cos^2\beta + \cos^2\gamma = 1.$$

EXAMPLES. 1. Two of the direction-angles of a straight line are 60° and 45°. Show that the third is 60°.

2. Find the distances of $(4, -7, 4)$ from the axes, and from the origin.

3. Find the distance of $(4, -2, -1)$ from $(6, 3, 2)$.

142. *To find the angle between two straight lines whose direction-cosines are given.*

Let OP, OQ, be parallels to the given lines through the origin, and let α, β, γ, and α', β', γ', be the angles made by OP and OQ, respectively, with the axes of X, Y, and Z. These angles

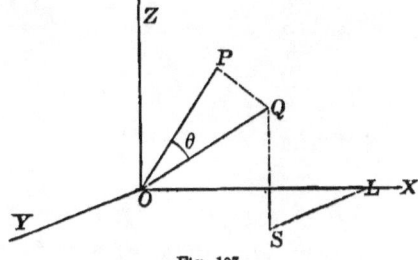

Fig. 107.

are, then, the direction-angles of the given lines. Let $QOP = \theta$, and x, y, z, be the coordinates of Q. Then (Art. 134), the projection of OQ upon OP is equal to the algebraic sum of the projections of OL, LS and SQ, upon OP. But the projection of OL on OP is (Art. 133)

$$OL \cos\alpha = OQ \cos\alpha' \cdot \cos\alpha.$$

Similarly, the projections of LS and SQ on OP are

$$LS \cos \beta = OQ \cos \beta' \cdot \cos \beta,$$

$$SQ \cos \gamma = OQ \cos \gamma' \cdot \cos \gamma.$$

The projection of OQ on OP is $OQ \cos \theta$. Hence

$$OQ \cos \theta = OQ \cos a' \cos a + OQ \cos \beta' \cos \beta + OQ \cos \gamma' \cos \gamma,$$

or $\qquad \cos \theta = \cos a' \cos a + \cos \beta \cos \beta' + \cos \gamma \cos \gamma'$,

or the cosine of the angle included between any two straight lines is the sum of the rectangles of their corresponding direction-cosines.

SECTION XVI. — THE PLANE.

143. General equation of a surface.

We have seen that any point (x, y, z) may be constructed by first locating the point (x, y) in the plane XY, and then laying off, on a perpendicular through this point to XY, the distance z. Hence, if z be made equal to any constant, as a, x and y remaining variables, the point (x, y, a) will lie in a plane parallel to XY. If, therefore,

$$f(x, y, z) = 0 \qquad (1)$$

be any equation between x, y, and z, and in this equation $z = a$, a constant, then $f(x, y, a) = 0$, being an equation between *two* variables, will represent a *line*, all of whose points are in a plane parallel to XY at a distance a from it. Similarly if $z = b$, $f(x, y, b) = 0$ will be the equation of a line in a plane parallel to XY at a distance b from it. Giving thus, successively, to z, all possible values, *i.e.*, letting z vary continuously between the limits assigned by the equation $f(x, y, z) = 0$, we obtain a series of lines, all of which are plane curves parallel to XY, which, taken together, form a surface of which $f(x, y, z) = 0$ is the equation. Hence $f(x, y, z) = 0$ *is the equation of a surface.*

To illustrate: let P be any point in space subject to the condition that the sum of the squares of its coordinates is a constant, or

$$x^2 + y^2 + z^2 = R^2. \qquad (2)$$

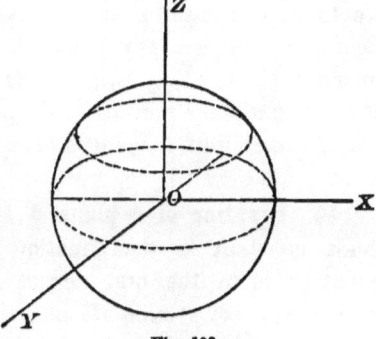

Fig. 108.

Since this sum is the square of the distance of P from the origin (Art. 137), it is evident that P is restricted to the surface of a sphere whose radius is R and whose centre is at the origin, and of which (2) is the equation. If we assume $z = 0$, then $x^2 + y^2 = R^2$ is the equation of a circle in the plane XY, whose radius is that of the sphere; that is, it is the great circle cut from the sphere by XY. If we make $z = a$, we have $x^2 + y^2 = R^2 - a^2$, which is also the equation of a circle, namely, that cut from the sphere by a plane parallel to XY at a distance from it equal to a. As a increases, the radius of the circle, $\sqrt{R^2 - a^2}$, diminishes, and when z is made equal to $a = R$, we have

$$x^2 + y^2 = 0, \text{ or } x = 0, \ y = 0,$$

the plane then touching the sphere at its highest point $(0, 0, R)$. z cannot be made greater than R, for then $x^2 + y^2 = R^2 - a^2$ would be impossible, since the sum of two squares cannot be negative, showing that no plane at a greater distance from XY than R can cut the surface of the sphere.

The lines cut from any surface by a plane are called **sections** of the surface. If x were made constant in (2), then

$$y^2 + z^2 = R^2 - a^2$$

would be the section cut by a plane parallel to YZ from the surface of the sphere; and if y were made constant, we should have the section made by a plane parallel to ZX, all of which would in this case be circles. And, in general, *if in the equation of any surface, $f(x, y, z) = 0$, one of the variables be made constant, the resulting equation is that of the line cut from the surface by a plane parallel to the plane of the other two axes and at a distance from it equal to the value assigned.*

144. Equation of a plane. If, when either x, y, or z is made constant in the equation $f(x, y, z) = 0$, the resulting equation is of the first degree between the two remaining variables, every section of the surface by planes parallel to the coordinate planes is a straight line, and the surface must be a plane. But this can be the case only when $f(x, y, z) = 0$ is of the first degree with respect to *all three* of the variables. Hence, *every equation of the form*

$$Ax + By + Cz + F = 0 \qquad (1)$$

is the equation of a plane.

145. Intercept form of the equation of a plane.

If in the equation of a plane,

$$Ax + By + Cz + F = 0, \qquad (1)$$

we make $y = z = 0$, we obtain $x = OQ = -\dfrac{F}{A}$, the intercept of the plane on X. Making $z = x = 0$, we have $y = OR = -\dfrac{F}{B}$, the intercept on Y; and for $x = y = 0$, $z = OS = -\dfrac{F}{C}$, the intercept on Z. Representing these intercepts by a, b, and c, respectively, we have

$$a = -\frac{F}{A}, \quad b = -\frac{F}{B}, \quad c = -\frac{F}{C},$$

whence

$$A = -\frac{F}{a}, \quad B = -\frac{F}{b}, \quad C = -\frac{F}{c}.$$

Substituting these values in (1), we have

$$\frac{x}{a} + \frac{y}{b} + \frac{z}{c} = 1, \qquad (2)$$

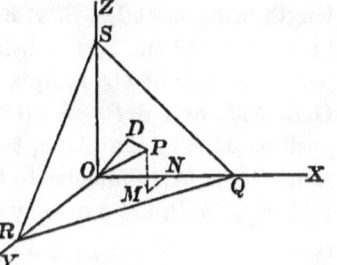

Fig. 109.

the equation of a plane in terms of its intercepts.

This form is not applicable when the plane passes through the origin; for in this case, since the origin is a point of the plane, $(0, 0, 0)$ must satisfy its equation, and from (1), $F = 0$, and $a = b = c = 0$.

To put the equation of a plane in the intercept form, transpose the absolute term to the second member, and, by division, make the second member positive unity. Thus, the intercept form of $3x - 6y + 2z - 6 = 0$ is $\dfrac{x}{2} - y + \dfrac{z}{3} = 1$, the intercepts being 2, -1, 3.

To write the equation of a plane whose intercepts are given, substitute their values in (2). Thus, the equation of the plane whose intercepts are 4, -3, -6, is

$$\frac{x}{4} - \frac{y}{3} - \frac{z}{6} = 1, \quad \text{or} \quad 3x - 4y - 2z - 12 = 0.$$

EXAMPLES. 1. Write the equation of a plane whose intercepts are 2, 6, 4; also -2, -3, 1.

Ans. $6x + 2y + 3z - 12 = 0$; $3x + 2y - 6z + 6 = 0$.

2. Put the following equations under the intercept form:

$$3x + 5y - z + 15 = 0, \quad x - y - z - 1 = 0.$$

3. Determine the intercepts of the planes of Ex. 2, without putting the equations under the intercept form.

146. Normal equation of the plane. Let QRS, Fig. 109, be any plane; OD a perpendicular upon it from the origin, its length being p, and α, β, γ, its direction-cosines. Let P be any point of the plane, x, y, z, being its coordinates. Then the projection on OD of OP is equal to the sum of the projections of ON, NM, and MP, on OD (Art. 134). But, whatever the position of P in the plane, the projection of OP on OD is p, since OD is perpendicular to the plane; and the projections of ON, NM, MP, are $x \cos \alpha$, $y \cos \beta$, $z \cos \gamma$ (Art. 133).

Hence $\qquad x \cos \alpha + y \cos \beta + z \cos \gamma = p \qquad$ (1)

is the normal equation of the plane, in which p is always positive. Since the sum of the squares of the direction-cosines of any line is unity, to put the equation of a plane under the normal form, we must introduce a factor R fulfilling the condition

$$(RA)^2 + (RB)^2 + (RC)^2 = 1,$$

in which A, B, C, are the coefficients of x, y, and z, in the given equation; hence

$$R = \frac{1}{\sqrt{A^2 + B^2 + C^2}}.$$

THE PLANE.

Thus, the normal form of $3x + 2y - z + 1 = 0$ is

$$-\frac{3x}{\sqrt{14}} - \frac{2y}{\sqrt{14}} + \frac{z}{\sqrt{14}} = \frac{1}{\sqrt{14}},$$

the second member being made positive; in which

$$-\frac{3}{\sqrt{14}}, \quad -\frac{2}{\sqrt{14}}, \quad \frac{1}{\sqrt{14}},$$

are the direction-cosines, and $p = \dfrac{1}{\sqrt{14}} =$ distance of the plane from the origin.

EXAMPLES. 1. Put the equations $3x + 5y - z + 15 = 0$, $x - y - z - 1 = 0$, under the normal form.

2. Find the distance of the plane $x - y + z - 1 = 0$ from the origin. In what angle is the perpendicular from the origin on the plane? *Ans.* $\dfrac{1}{\sqrt{3}}$; *in the fourth angle.*

3. Show that $4x + 7y + 4z - 9 = 0$ is at a distance unity from the origin.

4. Write the equation of a plane whose distance from the origin is 10, the direction-cosines of the perpendicular being $\dfrac{1}{2}, \dfrac{1}{3}, \dfrac{\sqrt{23}}{6}$. *Ans.* $3x + 2y + \sqrt{23}\,z - 60 = 0$.

5. Are the direction-cosines of Ex. 4 chosen at random?

6. Write the equation of a plane parallel to YZ at a distance from YZ equal to 6.

Since the plane is parallel to YZ, its intercepts on Y and Z are both infinity. Hence, from Eq. 2, Art. 113, $b = \infty$, $c = \infty$, and $x = a = 6$. Or, from Eq. 1, Art. 114, $a = 0$, $\beta = \gamma = 90°$; hence $x = p = 6$.

7. Write the equation of a plane parallel to X.

8. What are the equations of the coordinate planes?

147. *To write the equation of a plane through three given points.* Assuming the general equation

$$Ax + By + Cz + F = 0, \tag{1}$$

210   ANALYTIC GEOMETRY.

dividing by any one of the four constants, as F, and denoting the resulting coefficients by A', B', C', we obtain

$$A'x + B'y + C'z + 1 = 0. \qquad (2)$$

Substituting in this equation the coordinates of the three given points in succession, there results three equations between A', B', and C', from which the values of these latter may be determined. Substituting these values in (2), we have the equation required.

EXAMPLES. Write the equations of the planes through the following points.

$(1, 0, -2)$, $(3, 2, -1)$, $(5, -1, 2)$. *Ans.* $9x - 4y - 10z - 29 = 0$.
$(1, 2, 3)$, $(4, 5, 6)$, $(-7, 8, 9)$. *Ans.* $y - z + 1 = 0$.
$(0, 0, 0)$, $(1, 2, 4)$, $(1, -2, 6)$. *Ans.* $10x - y - 2z = 0$.
$(2, 0, 0)$, $(0, 2, 0)$, $(0, 0, 2)$. *Ans.* $x + y + z - 2 = 0$.
$(0, 1, 2)$, $(0, 2, 4)$, $(1, 0, 2)$. *Ans.* $2x + 2y - z = 0$.

148. *To find the angle between two given planes.*

The angle between the planes is the same as that between the perpendiculars upon the planes from the origin. Hence, if α, β, γ, and α', β', γ', are the direction-angles of the perpendiculars, and θ the angle between the latter, the required angle is given by the relation (Art. 142)

$$\cos\theta = \cos\alpha\cos\alpha' + \cos\beta\cos\beta' + \cos\gamma\cos\gamma'. \qquad (1)$$

If the equations of the planes are given in the normal form, we have only to substitute in (1) the coefficients of the variables, they being the direction-cosines (Art. 146). If the equations are in the general form

$$Ax + By + Cz + F = 0, \; A'x + B'y + C'z + F' = 0,$$

their normal forms will be

$$\frac{Ax + By + Cz}{\sqrt{A^2 + B^2 + C^2}} = \frac{F}{\sqrt{A^2 + B^2 + C^2}},$$

$$\frac{A'x + B'y + C'z}{\sqrt{A'^2 + B'^2 + C'^2}} = \frac{F'}{\sqrt{A'^2 + B'^2 + C'^2}},$$

THE PLANE. 211

in which the radicals have the opposite signs of the absolute terms in order to make the second members positive; and the direction-cosines are, respectively,

$$\left.\begin{array}{c}\dfrac{A}{\sqrt{A^2+B^2+C^2}},\ \dfrac{B}{\sqrt{A^2+B^2+C^2}},\ \dfrac{C}{\sqrt{A^2+B^2+C^2}},\\ \dfrac{A'}{\sqrt{A'^2+B'^2+C'^2}},\ \dfrac{B'}{\sqrt{A'^2+B'^2+C'^2}},\ \dfrac{C'}{\sqrt{A'^2+B'^2+C'^2}}.\end{array}\right\} \quad (2)$$

Substituting these values in (1),

$$\cos\theta = \frac{AA' + BB' + CC'}{\sqrt{A^2+B^2+C^2}\ \sqrt{A'^2+B'^2+C'^2}}. \quad (3)$$

Cor. 1. If the given planes are perpendicular to each other, $\theta = 90°$, $\cos\theta = 0$; hence *the condition of perpendicularity* is

$$AA' + BB' + CC' = 0, \quad (4)$$

or *the sums of the rectangles of the coefficients of the corresponding coordinates in the equations of the planes must be zero.*

Cor. 2. If the planes are parallel, the direction-cosines of their perpendiculars with respect to each axis must be equal. Hence, from (2)

$$\frac{A}{A'} = \frac{B}{B'} = \frac{C}{C'}, \quad (5)$$

since each ratio is equal to $\dfrac{\sqrt{A^2+B^2+C^2}}{\sqrt{A'^2+B'^2+C'^2}}$.

Hence *the condition of parallelism* is that *the ratios of the coefficients of the corresponding coordinates in the equations of the planes must be equal.*

EXAMPLES. 1. Find the angle between the planes $x + 2y - 2z + 1 = 0$ and $3x + 6y - 6z - 5 = 0$. *Ans.* $0°$.

2. Find the angle between the planes $2x + 2y + z + 1 = 0$, $4x - 4y + 7z - 1 = 0$. *Ans.* $\theta = \cos^{-1}\frac{7}{27}$.

3. Show that $3\left(x + y - \dfrac{z}{4}\right) + 1 = 0$ and $12(x+y) - 3z + 10 = 0$ are parallel.

4. Show that $x + 2y - 2z + 1 = 0$ is perpendicular to $2x + 5y + 6z - 11 = 0$; also $x + 2y + 3z + 1 = 0$ to $3x + 6y - 5z - 3 = 0$.

5. Write the equation of a plane parallel to $3x + 4y - z + 6 = 0$.

6. Write the equation of a plane perpendicular to $3x + 4y + z - 1 = 0$.

7. Find the distance between the parallel planes $x + 2y - 2z + 1 = 0$, $3x + 6y - 6z - 25 = 0$. *Ans.* $\frac{28}{9}$.

8. Prove that $Ax + By + Cz + F + k(A'x + B'y + C'z + F') = 0$ is the equation of a plane through the intersection of the planes
$$Ax + By + Cz + F = 0 \text{ and } A'x + B'y + C'z + F' = 0.$$
See Art. 37.

9. Write the equation of any plane through the intersection of $2x + 5y + z - 1 = 0$ and $x - y + z + 2 = 0$.

10. Explain how to determine k in Ex. 8 so that the plane shall pass through a given point. See Art. 37.

11. Write the equation of a plane through the intersection of $2x + y - z + 1 = 0$ and $3x + 4y + 2z + 6 = 0$, passing also through the point $(1, 1, 2)$. *Ans.* $28x + 9y - 21z + 5 = 0$.

12. Write the equation of a plane through the intersection of the planes of Ex. 11, and also passing through the origin.
Ans. $9x + 2y - 8z = 0$.

13. Prove that the distance from the point (x', y', z') to the plane $x \cos \alpha + y \cos \beta + z \cos \gamma = p$, is
$$x' \cos \alpha + y' \cos \beta + z' \cos \gamma - p,$$
or $\dfrac{Ax' + Bx' + Cz' - F}{\sqrt{A^2 + B^2 + C^2}}.$ See Art. 38.

14. Find the distances from the plane $5x + 2y - 7z + 9 = 0$ of the points $(1, -1, 3)$ and $(3, 3, 3)$.

15. Find the equation of a plane through $(1, 10, -2)$, parallel to the plane $2x + y - z + 6 = 0$. *Ans.* $2x + y - z - 14 = 0$.

16. Find the equation of a plane through $(1, -1, 3)$ perpendicular to the plane $2x + y - z + 6 = 0$.

17. Find the distance from $(8, 14, 8)$ to $4x + 7y + 4z - 18 = 0$. *Ans.* 16.

149. Traces of a plane. If in the equation of a plane, $Ax + By + Cz + F = 0$, we make $z = 0$, the resulting equation

$$Ax + By + F = 0 \tag{1}$$

applies to all points of the plane in XY, and is therefore the equation of RQ (Fig. 109), the intersection of the plane with XY. For like reasons

$$By + Cz + F = 0,$$
$$Ax + Cz + F = 0,$$

are the equations of the intersections RS and SQ. These intersections are called the **traces** of the plane. Solving (1) for y, we have

$$y = -\frac{A}{B}x - \frac{F}{B},$$

and the corresponding trace for any other plane would be

$$y = -\frac{A'}{B'}x - \frac{F'}{B'}.$$

If the traces are parallel, $\dfrac{A}{B} = \dfrac{A'}{B'}$, or $\dfrac{A}{A'} = \dfrac{B}{B'}$. If the corresponding traces on the other coordinate planes are also parallel, in which case the planes themselves are parallel, we obtain in like manner

$$\frac{A}{A'} = \frac{B}{B'} = \frac{C}{C'},$$

the condition already found.

SECTION XVII.—THE STRAIGHT LINE.

150. Equations of the straight line.

Assuming the equation of a plane in the intercept form,

$$\frac{x}{a} + \frac{y}{b} + \frac{z}{c} = 1,$$

if we impose the condition that the plane shall be perpendicular to XZ, its Y-intercept b will be infinity, and its equation assumes the form

$$Ax + Cz + F = 0, \qquad (1)$$

whatever the value of y. Hence every equation of the first degree between two variables represents a plane perpendicular to the corresponding coordinate plane, the third variable being indeterminate. Therefore

$$B'y + C'z + F' = 0,$$

x being indeterminate, represents a plane perpendicular to YZ.

Let $ABDL$ be the plane represented by (1), and $AHDC$ that represented by (2). Values of x, y, z, which satisfy both (1) and (2) locate a point in both planes, that is, on AD, their line of intersection. Hence, while taken separately (1) and (2) are equations of planes perpendicular respectively to XZ and YZ, if taken together they represent a straight line in space. Thus, let x be the independent variable, and any value as $x = OM$ be substituted in (1). From (1) we may then find $z = MS$ and locate the point S in XZ. Now so long as (1) is considered independent of (2), it represents the plane $LABD$, and the value of y may be assumed at pleasure. But if (1) and (2) are simultaneous, y must be derived from (2) after the value $z = MS$ has been substituted in it. Since this value of y satisfies (2), $y = SP$ must locate a point P in the plane $AHDC$, and P must lie on the intersection AD.

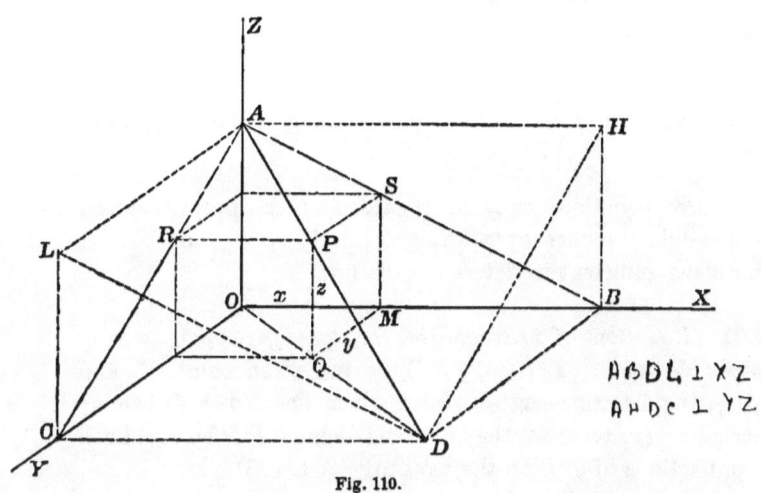

Fig. 110.

The line AD is evidently completely determined by (1) and (2), since the planes can intersect in but one straight line.

Since (1) is true for all values of y, it is true for $y = 0$, and is then the equation of the trace AB of $ABDL$ on XZ. Hence (1) is the equation of the plane $ABDL$, or of its trace AB, according as $y = \frac{0}{0}$ or $y = 0$. Similarly (2) is the equation of the plane $AHDC$, or of its trace AC, according as $x = \frac{0}{0}$ or $x = 0$. But AB and AC are the projections of AD on XZ and YZ, since the planes $ABDL$ and $AHDC$ are perpendicular respectively to XZ and YZ. Hence the straight line AD is determined when its projections on any two coordinate planes are given.

Eliminating z between (1) and (2), we have an equation of the form

$$A''x + B''y + F'' = 0, \qquad (3)$$

which, in like manner, is the equation of a plane perpendicular to XY, or of its trace on XY, according as we regard $z = \frac{0}{0}$ or $z = 0$. This plane is evidently the plane AOD, passing through the intersection AD of (1) and (2), since (1), (2), and (3) are simultaneous; and its trace is OD, the projection of AD on XY.

Hence, in general, if we assume any two equations of the first degree between two variables, as

$$f(x, z) = 0, \quad f'(y, z) = 0,$$

and eliminate the common variable, obtaining the third equation,

$$f''(x, y) = 0,$$

these three equations may be regarded as the projections on the coordinate planes of a straight line in space, any two of them being sufficient to determine the line.

151. *Equations of a straight line through a given point having a given direction.* Let (x', y', z') be the given point, P', and α, β, γ, the direction-angles of the given line. Let P be any other point (x, y, z) of the line, and denote $P'P$ by r. Then the projections of $P'P$ on the axes are (Art. 133)

$$x - x' = r\cos\alpha, \quad y - y' = r\cos\beta, \quad z - z' = r\cos\gamma;$$

from which
$$\frac{x - x'}{\cos\alpha} = \frac{y - y'}{\cos\beta} = \frac{z - z'}{\cos\gamma}, \tag{1}$$

which are the required equations, any two being sufficient to determine the line.

Eq. (1) is called the **symmetrical** form.

152. *To put the equations of a straight line under the symmetrical form.*

The symmetrical form being

$$\frac{x - x'}{\cos\alpha} = \frac{y - y'}{\cos\beta} = \frac{z - z'}{\cos\gamma},$$

the condition that the equations of a straight line are in this form is that the sum of the squares of the denominators is unity. Let

$$\frac{x - x'}{L} = \frac{y - y'}{M} = \frac{z - z'}{N}$$

be the given equations. Dividing the denominators by $\sqrt{L^2+M^2+N^2}$, we have

$$\frac{x-x'}{\dfrac{L}{\sqrt{L^2+M^2+N^2}}} = \frac{y-y'}{\dfrac{M}{\sqrt{L^2+M^2+N^2}}} = \frac{z-z'}{\dfrac{N}{\sqrt{L^2+M^2+N^2}}},$$

in which the sum of the squares of the denominators $= 1$. Thus, let $3x-2z+1=0$, $4x-y=0$, be the given line. Then

$$\frac{x}{1} = \frac{2z-1}{3} = \frac{y}{4}.$$

Dividing the denominators by $\sqrt{1^2+3^2+4^2} = \sqrt{26}$,

$$\frac{x}{\dfrac{1}{\sqrt{26}}} = \frac{2z-1}{\dfrac{3}{\sqrt{26}}} = \frac{y}{\dfrac{4}{\sqrt{26}}},$$

the direction-cosines being $\dfrac{1}{\sqrt{26}}$, $\dfrac{3}{\sqrt{26}}$, $\dfrac{4}{\sqrt{26}}$.

EXAMPLES. 1. Find the equations of the intersection of $x-y+z-2=0$, and $x+y+2z-1=0$, and determine the position of the line.

Eliminating y and z in succession, we have

$$2x+3z-3=0, \quad x-3y-3=0,$$

or

$$\frac{x}{1} = \frac{z-1}{\dfrac{-2}{3}} = \frac{y+1}{\dfrac{1}{3}},$$

whence the line passes through $(0, -1, 1)$, and its direction-cosines are

$$\frac{3}{\sqrt{14}}, \quad -\frac{2}{\sqrt{14}}, \quad \frac{1}{\sqrt{14}}.$$

2. Find the intersection of $x+y-z+1=0$ and $4x+y-2z+2=0$.

Ans. *A line through* $(0, 0, 1)$, *whose direction-cosines are*

$$\frac{1}{\sqrt{14}}, \quad \frac{2}{\sqrt{14}}, \quad \frac{3}{\sqrt{14}}.$$

3. Determine the position of $x = 4z + 3$, $y = 3z - 2$.

Ans. A line through $(3, -2, 0)$, *whose direction-cosines are*
$$\frac{4}{\sqrt{26}}, \frac{3}{\sqrt{26}}, \frac{1}{\sqrt{26}}.$$

4. Write the equation of a line through $(1, 2, -6)$, having $\frac{2}{3}, \frac{1}{3}, \frac{2}{3}$, for direction-cosines.

Ans. $x - 2y + 3 = 0$, $2y - z - 10 = 0$.

5. Write the equation of a line through $(1, 4, -3)$ parallel to Z. *Ans.* $x = 1$, $y = 4$.

153. *Equations of a straight line through two given points.*

Let (x', y', z'), (x'', y'', z''), be the given points. The equations of a straight line through (x', y', z') are

$$\frac{x - x'}{\cos a} = \frac{y - y'}{\cos \beta} = \frac{z - z'}{\cos \gamma}. \tag{1}$$

Since the point (x'', y'', z'') is also on the line, its coordinates must satisfy (1); hence

$$\frac{x'' - x'}{\cos a} = \frac{y'' - y'}{\cos \beta} = \frac{z'' - z'}{\cos \gamma}. \tag{2}$$

Dividing (1) by (2), member by member,

$$\frac{x - x'}{x'' - x'} = \frac{y - y'}{y'' - y'} = \frac{z - z'}{z'' - z'}, \tag{3}$$

which are the required equations, any two of which determine the line.

EXAMPLES. 1. Write the equations of the straight line passing through $(1, 2, 4)$, $(-3, 6, -1)$.

Ans. $\dfrac{x-1}{-4} = \dfrac{y-2}{4} = \dfrac{z-4}{-5}$.

2. Find the direction of the line of Ex. 1.

3. Find the points in which the line of Ex. 1 pierces the coordinate planes. *Ans. The line pierces* XY *in* $(-\frac{11}{5}, \frac{26}{5})$.

THE STRAIGHT LINE. 219

4. Write the equations of lines through the following points, and find their directions:

$(2, 1, -1), (-3, -1, 1)$; $(6, 2, 4), (-6, -3, 1)$.

5. A line passes through $(1, 1, 2)$ and the origin; find its equations.

6. A line passes through $(1, 6, 3,)$ $(1, -6, 2)$. Find the equations of its projections on the coordinate planes.

154. *To find the angle between two given straight lines.*

Let
$$\frac{x-x'}{L'} = \frac{y-y'}{M'} = \frac{z-z'}{N'},$$

$$\frac{x-x''}{L''} = \frac{y-y''}{M''} = \frac{z-z''}{N''},$$

be the given lines. The angle between the two lines is given by the relation (Art. 142)

$$\cos\theta = \cos\alpha'\cos\alpha'' + \cos\beta'\cos\beta'' + \cos\gamma'\cos\gamma'',$$

in which α', β', γ', and $\alpha'', \beta'', \gamma''$, are their direction-angles. But (Art. 152),

$$\left.\begin{array}{l} \cos\alpha' = \dfrac{L'}{\sqrt{L'^2 + M'^2 + N'^2}}, \\[6pt] \cos\beta' = \dfrac{M'}{\sqrt{L'^2 + M'^2 + N'^2}}, \\[6pt] \cos\gamma' = \dfrac{N'}{\sqrt{L'^2 + M'^2 + N'^2}}, \end{array}\right\} \quad (1)$$

$$\left.\begin{array}{l} \cos\alpha'' = \dfrac{L''}{\sqrt{L''^2 + M''^2 + N''^2}}, \\[6pt] \cos\beta'' = \dfrac{M''}{\sqrt{L''^2 + M''^2 + N''^2}}, \\[6pt] \cos\gamma'' = \dfrac{N''}{\sqrt{L''^2 + M''^2 + N''^2}}, \end{array}\right\} \quad (2)$$

220   ANALYTIC GEOMETRY.

Hence $\quad \cos\theta = \dfrac{L'L'' + M'M'' + N'N''}{\sqrt{L'^2 + M'^2 + N'^2}\sqrt{L''^2 + M''^2 + N''^2}}.$ (3)

Cor. 1. If the lines are parallel, the corresponding direction-cosines are equal, each to each; hence from (1) and (2),

$$\frac{L'}{L''} = \frac{M'}{M''} = \frac{N'}{N''}$$

are *the conditions of parallelism.*

Cor. 2. If the lines are perpendicular to each other,

$$\theta = 90°, \quad \cos\theta = 0;$$

hence $\quad\quad L'L'' + M'M'' + N'N'' = 0$

is *the condition of perpendicularity.*

Examples. 1. Find the equations of the sides of the triangle whose vertices are (1, 2, 3), (3, 2, 1), (2, 3, 1), and the angles of the triangle.

Ans. $\left\{\begin{array}{l}x+z=4\\y=2\end{array}\right\}, \left\{\begin{array}{l}x+y=5\\z=1\end{array}\right\}, \left\{\begin{array}{l}x-y=-1\\2x+z=5\end{array}\right\}, \dfrac{\pi}{6}, \dfrac{\pi}{2}, \dfrac{\pi}{3}.$

2. Find the angle between

$$y = 5x + 3, \quad z = 3x + 5, \quad \text{and} \quad y = 2x + 1, \quad z = x.$$

3. Show that

$$4x - 3y - 10 = 0, \quad y + 4z + 26 = 0,$$

and $\quad\quad 7x - 2y + 26 = 0, \quad 34y + 7z - 90 = 0,$

are perpendicular to each other.

4. Show that $\quad x = 2z + 1, \quad y = 3z + 4,$

and $\quad\quad x = 3 - 2z, \quad y = z - 2,$

are perpendicular to each other.

5. Show that

$$2x - y + 1 = 0, \quad 3y - 2z + 5 = 0,$$

and $\quad\quad 2x - y - 7 = 0, \quad 3y - 2z + 7 = 0,$ are parallel.

6. Find the conditions that the straight line

$$\frac{x-x'}{L} = \frac{y-y'}{M} = \frac{z-z'}{N}$$

is parallel or perpendicular to the plane

$$Ax + By + Cz + F = 0.$$

The line is parallel to the plane when it is perpendicular to the perpendicular on the plane. But the direction-cosines of the perpendicular are proportional to A, B, C (Art. 146), and the direction-cosines of the line are proportional to L, M, N (Art. 152). Hence the condition of parallelism is (Art. 154)
$$AL + BM + CN = 0.$$

The line is perpendicular to the plane when it is parallel to the perpendicular on the plane; hence, the condition of perpendicularity is

$$\frac{L}{A} = \frac{M}{B} = \frac{N}{C}.$$

7. Find the equation of a line through $(-2, 3, 5)$ perpendicular to $2x + 8y - z - 4 = 0$.

Ans. $x + 2z - 8 = 0,\ y + 8z - 43 = 0.$

8. Show that $2x - y = 0,\ 3y - 2z = 0$ is perpendicular to

$$x + 2y + 3z - 6 = 0.$$

9. Show that $z = 3,\ x + y = 3$ is parallel to

$$x + y + z - 6 = 0,$$

and to the trace of the latter on XY.

CHAPTER VI.

SURFACES OF REVOLUTION, CONIC SECTIONS, AND HELIX.

SECTION XVIII.—SURFACES OF REVOLUTION.

155. Defs. *A line is said to be revolved about a straight line as an axis when every point of the line describes a circle whose centre is in the axis and whose plane is perpendicular to the axis.*

The moving line is called the **generator**, and the surface which it generates a **surface of revolution**. It follows from the definition of revolution that every plane section of a surface of revolution perpendicular to the axis is a circle, and that every plane section through the axis is the generator in some one of its positions. A plane through the axis is called a **meridian plane**, and the section cut by such a plane is called a **meridian**.

156. General equation of a surface of revolution.

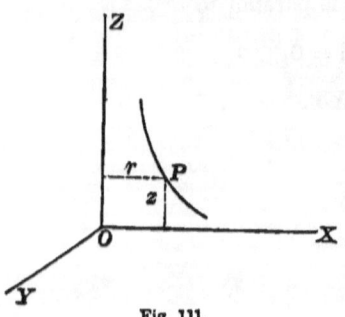

Fig. 111.

Let the axis of Z be the axis of revolution, the generator a plane curve whose initial position is in the plane XZ, and P any point of the generator. Since the generator is in the plane XZ, its equation will be $x = f(z)$, but as it revolves about Z, the x-coordinate of any point as P will differ from that of its initial position. Hence, to distinguish the x-coordinate of the surface from that of the generator in its initial position, represent the latter by r; then the equation of the generator will be

$$r = f(z). \qquad (1)$$

But P remains at the same distance, r, from Z during the revolution; hence (Art. 137)

$$r^2 = x^2 + y^2. \qquad (2)$$

Substituting in (2) the value of r from (1),

$$x^2 + y^2 = [f(z)]^2, \qquad (3)$$

is the general equation of a surface of revolution. In any particular case substitute in (3) $f(z)$ from the equation of the generator.

157. The sphere. If a circle be revolved about any one of its diameters the surface generated will be a sphere. Let the diameter coincide with Z and the centre with the origin. Then the equation of the generator will be

$$r^2 + z^2 = R^2,$$

whence $r^2 = [f(z)]^2 = R^2 - z^2$. Substituting this in the general equation $x^2 + y^2 = [f(z)]^2$, we have

$$x^2 + y^2 + z^2 = R^2,$$

which is the required equation.

158. The prolate spheroid, or ellipsoid. This is the surface generated by the revolution of an ellipse about the transverse axis. Let the transverse axis coincide with Z and the centre with the origin. Then the equation of the generator is

$$a^2 r^2 + b^2 z^2 = a^2 b^2,$$

whence $$r^2 = \frac{b^2}{a^2}(a^2 - z^2) = [f(z)]^2.$$

Substituting this in $x^2 + y^2 = [f(z)]^2$, we have

$$a^2(x^2 + y^2) + b^2 z^2 = a^2 b^2, \quad \text{or} \quad \frac{x^2}{b^2} + \frac{y^2}{b^2} + \frac{z^2}{a^2} = 1. \quad (1)$$

If $a^2 = b^2 = R^2$, the ellipsoid becomes a sphere. By definition, plane sections parallel to XY are circles. Let the student prove that plane sections parallel to XZ and YZ are ellipses.

159. The oblate spheroid, or ellipsoid. This is the surface generated by the revolution of the ellipse about the conjugate axis. Let the conjugate axis coincide with Z and the centre with the origin. Then the equation of the generator is

$$a^2 z^2 + b^2 r^2 = a^2 b^2,$$

whence $\quad r^2 = \dfrac{a^2}{b^2}(b^2 - z^2) = [f(z)]^2,$

and $\quad b^2(x^2 + y^2) + a^2 z^2 = a^2 b^2, \quad \text{or} \quad \dfrac{x^2}{a^2} + \dfrac{y^2}{a^2} + \dfrac{z^2}{b^2} = 1,$

which is the required equation. If $a^2 = b^2 = R^2$, the ellipsoid becomes a sphere.

Let the student determine the plane sections parallel to the coordinate planes.

160. The paraboloid. This is the surface generated by the revolution of a parabola about its axis. Let the vertex of the parabola be at the origin, the axis coinciding with Z. Then the equation of the generator is

$$r^2 = 2pz = [f(z)]^2,$$

and the required equation is

$$x^2 + y^2 = 2pz. \quad (1)$$

Let the student show that plane sections parallel to YZ and XZ are parabolas.

161. The hyperboloid of two nappes. If an hyperbola be resolved about its transverse axis, the surface generated is

called the hyperboloid of two nappes. With the centre at the origin and the transverse axis coincident with Z, the equation of the generator is
$$a^2r^2 - b^2z^2 = -a^2b^2,$$
whence
$$a^2(x^2+y^2) - b^2z^2 = -a^2b^2, \qquad (1)$$
or
$$\frac{x^2}{b^2} + \frac{y^2}{b^2} - \frac{z^2}{a^2} = -1,$$
is the required equation.

Let the student determine the plane sections parallel to the coordinate planes.

162. Hyperboloid of one nappe. This is the surface generated by the revolution of an hyperbola about its conjugate axis. Assuming the centre at the origin and conjugate axis coincident with Z, the equation of the generator is
$$a^2z^2 - b^2r^2 = -a^2b^2,$$
and that of the surface is
$$b^2(x^2+y^2) - a^2z^2 = a^2b^2, \qquad (1)$$
or
$$\frac{x^2}{a^2} + \frac{y^2}{a^2} - \frac{z^2}{b^2} = 1.$$

Let the student determine the sections.

163. Cylinder of revolution. If a straight line revolve about another to which it is parallel, it will generate the surface of a circular cylinder. Let Z be the axis and $r = R$ the equation of the generator parallel to Z in the plane XZ. Then
$$r = R = f(z),$$
and
$$x^2 + y^2 = R^2 \qquad (1)$$
is the required equation, z being indeterminate.

Let the student show that sections parallel to z are two parallel straight lines, or one straight line, elements of the cylinder.

164. Cone of revolution. If a straight line revolves about another straight line which it intersects, the surface generated is that of a cone. Any position of the generator is called an **element** of the cone.

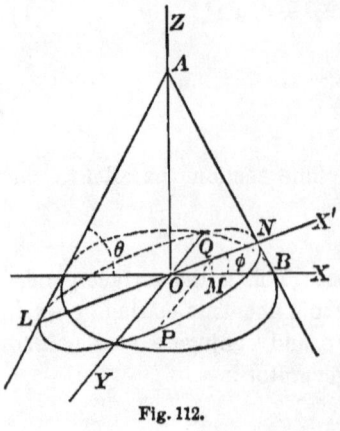

Fig. 112.

Let AB be the generator, and Z the axis of revolution. The cone will be a right cone whose vertex is A, $OA = h$ being the altitude and $OB = R$ the radius of the base in the plane XY. The coordinates of A and B are $(0, h)$, $(R, 0)$, and the equation of the generator $z = -\dfrac{h}{R} r + h$, whence

$$r^2 = [f(z)]^2 = \frac{R^2}{h^2}(h-z)^2, \qquad (1)$$

and the equation of the surface is

$$(x^2 + y^2)\frac{h^2}{R^2} = (h - z)^2, \qquad (2)$$

or if θ = angle which the generator makes with X = angle made by the elements of the cone with the plane of the base,

$$(x^2 + y^2)\tan^2\theta = (h - z)^2. \qquad (3)$$

If the vertex A is at an infinite distance, the cone becomes the cylinder. In this case $h = \infty$, and from (1)

$$[f(z)]^2 = \left[R^2 - \frac{2zR^2}{h} + \frac{z^2 R^2}{h^2}\right]_{h=\infty} = R^2,$$

and we obtain the equation of the cylinder $x^2 + y^2 = R^2$, as before.

Let the student prove that every plane section parallel to Z is an hyperbola.

SECTION XIX. — THE CONIC SECTIONS.

165. General equation. Let any plane (Fig. 112) be passed through the axis of Y, cutting the section LPN from the surface of the cone and the line LN from the plane ZX; and let $XON = \phi$, the inclination of the plane to XY. Since the cutting plane is perpendicular to ZX, its equation will be that of its trace LN, or

$$z = \tan \phi \cdot x.$$

To refer the curve of intersection LPN to axes in its own plane, let OY be the axis of Y, and $OX' = ON$ produced the new axis of X; then the coordinates of P referred to the primitive axes are $OM = x$, $MQ = z$, $QP = y$, and referred to the new axes are $x' = OQ$, $y' = QP$. Hence

$y = y'$, $x = OM = OQ \cos \phi = x' \cos \phi$, $z = MQ = OQ \sin \phi = x' \sin \phi$.

If these values, which are true for the point P common to both the plane and the cone, be substituted in Eq. (3) Art. 164, we shall have the equation of the plane section referred to the axes $X'OY$. Making those substitutions, and omitting the accents,

$$(x^2 \cos^2 \phi + y^2) \tan^2 \theta = (h - x \sin \phi)^2,$$

whence

$$y^2 \tan^2 \theta + x^2 (\cos^2 \phi \tan^2 \theta - \sin^2 \phi) + 2hx \sin \phi - h^2 = 0,$$

or, since $\sin^2 \phi = \cos^2 \phi \tan^2 \phi$,

$$y^2 \tan^2 \theta + x^2 \cos^2 \phi (\tan^2 \theta - \tan^2 \phi) + 2hx \sin \phi - h^2 = 0. \quad (1)$$

DISCUSSION OF THE EQUATION. Being of the second degree between x and y, this equation represents a conic.

If $\phi > \theta$, $\tan^2 \phi > \tan^2 \theta$, $B^2 - 4AC$ (Art. 80) is positive, and the section is an hyperbola.

If $\phi < \theta$, $\tan^2\phi < \tan^2\theta$, $B^2 - 4AC$ is negative, and the section is an ellipse.

If $\phi = \theta$, $\tan^2\phi = \tan^2\theta$, $B^2 - 4AC$ is zero, and the section is a parabola.

Hence the section is an hyperbola, ellipse, or parabola, according as the cutting plane makes an angle with the plane of the base greater than, less than, or equal to, that which the elements do.

If $h = 0$, the equation becomes

$$y^2 \tan^2\theta + x^2 \cos^2\phi (\tan^2\theta - \tan^2\phi) = 0,$$

and the plane passes through the vertex which is at the origin. In this case if $\phi > \theta$, the equation takes the form $y = \pm ax$, and represents two straight lines through the origin. If $\phi < \theta$, it is satisfied only for $x = 0$, $y = 0$, and represents a point. If $\phi = \theta$, it reduces to $y = 0$, the equation of X. These are particular cases of the hyperbola, ellipse, and parabola, respectively.

If $\phi = 0$; a particular case of $\phi < \theta$, the section is a circle by definition.

If $h = \infty$, the cone becomes a cylinder. Putting $h = \infty$ in Eq. (1), after substituting for $\tan^2\theta$ its value $\dfrac{h^2}{R^2}$ and dividing through by h^2, we have

$$y^2 + x^2 \cos^2\phi = R^2;$$

which is the equation of an ellipse, except when $\phi = 0°$ and $\phi = 90°$, in which cases the section is a circle, or two parallel straight lines, elements of the cylinder.

Having thus given to ϕ all possible values from $0°$ to $90°$, and h all possible values from 0 to infinity, we have found every section of the cone except those parallel to the **axis of revolution**, which latter have been already considered in Arts. 163 and 164. Thus every plane section is seen to be one of the varieties of the **conics**.

SECTION XX. — THE HELIX.

166. Defs. If a rectangular sheet of paper be rolled up into a right cylinder with a circular base, any straight line drawn on the paper, not parallel to its sides, will become a curve called the **helix**. Or it may be defined as the curve assumed by the hypothenuse of a right-angled triangle whose base is tangent to the base of the cylinder and whose plane is perpendicular to the radius of the base through the point of contact, when the triangle is wrapped around the cylinder. The helix forms the edge of the common screw. It follows from the definition that the helix makes a constant angle with the elements of the cylinder; namely, the acute angle at the base of the triangle.

167. Equations of the helix. Let the axis of the cylinder coincide with Z, $OA = R =$ radius of its base in the plane XY, P being any point of the helix, $a =$ constant angle at the base of the triangle, the vertex of this angle being assumed on the axis of X at A, and $\phi = AOQ =$ angle made by the projection of the radius vector OP on XY with X. Then

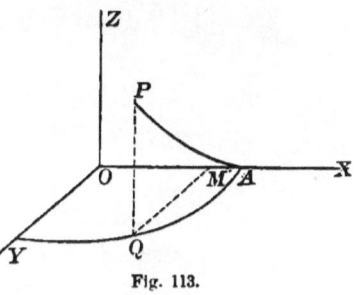

Fig. 113.

$x = OM = OQ \cos\phi = R\cos\phi, \; y = MQ = OQ \sin\phi = R\sin\phi,$

$z = QP =$ base of triangle $\times \tan a = QA \cdot \tan a = R\phi \tan a.$

Hence, if $k = \tan a$, the equations of the helix are

$$x = R\cos\phi, \; y = R\sin\phi, \; z = kR\phi.$$

www.ingramcontent.com/pod-product-compliance
Lightning Source LLC
Chambersburg PA
CBHW022007220426
43663CB00007B/989